BARRON'S

GMAT®
MATH WORKBOOK

ENDER MARKAL, MBA, CFA

BARRON'S

About the Author

Ender Markal is the founder of SFTutors, a San Francisco-based tutoring organization that helps students succeed on standardized tests and gain admission to top graduate schools. He has tutored math for almost 20 years, helping students overcome math anxiety and gain admission to top schools. He has a B.S. in mechanical engineering, an MBA in finance, and is a CFA.

All inquiries should be addressed to:
Barron's Educational Series, Inc.
250 Wireless Boulevard
Hauppauge, New York 11788
www.barronseduc.com

ISBN: 978-0-7641-4534-6

Library of Congress Control Number: 2010931205

Printed in the United States of America
9 8 7 6 5 4 3 2 1

CONTENTS

Getting Started

SECTION 1.1: LAYING THE FOUNDATION

Gaining admission to top business schools is becoming harder and more competitive each year. The 2008–09 test year, for example, set a record with more than 265,000 GMAT tests administered, representing an 8% increase over the previous year. Today, aspiring graduate school students compete with MBA hopefuls from all over the world for limited spots offered by top schools.

The importance of your GMAT score, which will strongly influence your acceptance into business programs, is contingent on a number of factors. If you are a recent graduate, the GMAT score is especially important because you lack the advantage of years of work experience that can set you apart among tens of thousands of candidates. For older applicants, the GMAT is particularly vital because admissions committees consider how long you've been out of school and will likely place more weight on your GMAT score if you have been out of school for some time.

Fortunately, boosting your GMAT score, as an integral part of your application, is still within your control. With appropriate preparation, you can use the GMAT test to your advantage and present a graduate school application that stands out among many others. The GMAT is your only chance to "make up" for weaker parts of your application and further enhance your application with a more recent and outstanding indication of your eligibility.

GMAT is *most* concerned with your ability to execute. This book is designed to introduce you to fundamental mathematical concepts contained within the GMAT. Because understanding math concepts theoretically is one thing, and their execution is another, this book presents a range of problems through which you can apply lessons that familiarize you with those principles, bridging the gap between basic math concepts and specific GMAT questions.

The comprehensive examples in this book are designed to ensure that you fully understand each concept before you proceed to self-practice. The extensive sample problems that accompany each section allow you to test your mastery of individual principles and your process as a whole.

While none of these concepts is any more challenging than what you have already learned in high school, you will see that connecting the dots between concepts and solutions might be difficult at first. The *good* news is that you will find many explanations in this book that will help you establish a direct line between basic concepts and solution applications. In addition, most of the concepts that will be new to you will be those that have to do with the *structure* of the test. Studying for this exam should be more a matter of refreshing your memory than it will be learning new concepts.

Your success will ultimately be the product of a proper GMAT preparation strategy. This book will show you how to strategically approach two different aspects of the GMAT: developing a preparatory system months and weeks before you take the exam, and mastering strategies you can implement on test day.

Subject explanations have been prepared with the busy test-taker in mind and organized by bullet points. There are reminder boxes throughout the book that will help you carry out solutions easier without the aid of a calculator.

We present you this valuable and practical guide to reach your life-changing goal. Now it is your turn to take control and accomplish it.

SECTION 1.2: WELCOME TO THE GMAT

The GMAT is an acronym for the Graduate Management Admission Test, a standardized exam administered internationally by the Graduate Management Admission Council (GMAC) and taken by students applying to business schools. Almost 2,000 business schools—nationally and internationally—consider the GMAT scores as integral components of candidates' admissions materials and accept the exam as one legitimate indicator of subsequent academic performance. Most of the testing locations around the world (and certainly all of the testing sites in North America) now administer the GMAT only via a computer program. Test takers no longer have the prerogative to opt for the pencil and paper format. For those of you who have just recently decided that business school is the appropriate next step or who have not taken a standardized test since the SAT (and you've been trying to forget *that* experience for a long time), you actually have very few reasons to feel any distress. The object of the test is to appraise your ability to reason methodically and to apply the mathematical and verbal skills you've accumulated throughout your career as a student and, if you've been lucky, in your employment since graduating. In essence, mastering this test is a matter of recollection and recall, as well as familiarizing yourself with both the format of the exam and the manner in which the questions are posed.

The GMAT: Not Your Nemesis

Many of you might be under the impression that this exam is going to be the most difficult hurdle to clear in your admissions process. This does not have to be so if you have allotted yourself a considerable amount of practice time and if you allow yourself to be guided by logic rather than by emotion during the course of your preparation. Although the GMAT is taken specifically by individuals hoping to enter business school, it does not endeavor to measure your comprehension of any particular business-related subject nor require you to have any business experience. (Schools might require you to have business experience, which is something to keep in mind when making application decisions.) You will be expected to have mastered the basics of "reading, writing, and 'rithmetic." The verbal segment will assume you understand the standard conventions of written English and that you are capable of presenting coherent and compelling arguments in two analytical essays. The math segment will assume you know the basics of algebra, geometry, and arithmetic and that you can apply these concepts in situations that are slightly more challenging than those you confronted in high school math class.

The GMAT essentially serves as an objective supplement to the more subjective gauges utilized in the candidate selection process, such as letters of reference, transcripts from previous institutions, and in-person interviews. Admissions offices understand that every applicant comes from a different background and a different accumulation of experiences—economically, socially, educationally, and so forth—than every other applicant. Thus the GMAT, business schools maintain, is a safe gauge by which applicants can be assessed under similar conditions and measured by the same standard. So you see, the GMAT is not necessarily your enemy. In most cases, it is an exam that simply requires you spend some time brushing off the cobwebs in those recesses of your brain where you've stored your knowledge of syntax and linear equations. (If you have completed college, you have undoubtedly been exposed to these skills before.) Half the battle of conquering the GMAT is simply knowing what to expect on the exam. By committing to working through this book, you are already on your way.

Your Scores and the Admissions Process

The influence your score has on your acceptance into business programs is contingent on a number of factors. Admissions committees will consider, for example, how long you've been out of school. They will likely place more weight on the GMAT scores of students who have been out of school for some time than on students who have recently graduated. Each institution you apply to is also likely to have its own estimate of acceptable and preferable GMAT scores. Many post averages of their entering classes on their websites. Be sure you check with the schools you are interested in so that you can set your goals. If, for example, your GPA is a little lower than the average GPA of a recent entering class, your goal should be to score a little higher than that class's mean GMAT score in order to compensate. Keep in mind, too, that other factors all carry weight, like your letters of recommendation, your work experience, and your essays. Be sure not to get so caught up in the GMAT that you fail to give each of these its requisite consideration.

SECTION 1.3: HOW TO USE THIS BOOK TO YOUR BEST ADVANTAGE

This book was designed not only to introduce you to the fundamental mathematical concepts. It was also designed to present you with a wide range of problems through which you can continue to familiarize yourself with those mathematical principles—through application. Understanding something on an abstract or theoretical level is one thing; execution is another. As you know, the GMAT is most concerned with your ability to execute. What we should pause here to stress, then, is how important it is that you work out even the sample problems, as opposed to merely reading the solutions. Reading is not going to help you absorb and retain nearly as well as will constant application.

The comprehensive examples in this book will help ensure you fully understand each concept before you proceed to self-practice. The extensive sample problems that accompany each section will allow you to test your mastery of individual principles and your process as a whole. Be sure to double-check the book once you've arrived at the solutions to these practice problems. We might be able to show you an easier or quicker way to arrive at that same answer (and quick and easy is how you will have to play this one).

Although none of these concepts is any more formidable than what you have already learned in your high school geometry and algebra classes, chances are high that you've been balancing your checkbook and estimating your taxes with a calculator ever since. The good news is that most of the concepts that will be new to you will be those that have to do with the structure of the test. The data sufficiency questions, for example, may seem a little daunting at first, but we'll talk more about those later. Studying for this exam should be more a matter of refreshing your memory than it will be learning new concepts. Furthermore, much of your success is going to be a product of strategy. As we all know, knowledge without systematic application tends to be knowledge ill used. This book, will focus on how to approach two different aspects of the GMAT strategically. It will help you develop a preparatory system for the months and weeks before you take the exam. It will also provide you with a number of strategies to implement on test day as you sit calmly and confidently before the screen. The mastery, so to speak, will be in the method.

Chapter Divisions

The math covered on the GMAT has been broken down into five chapters in this book. A chapter each is dedicated to arithmetic, algebra, and geometry. Chapter 4 gives special attention to word problems—those perplexing questions that actually test more than mathematical aptitude. In fact, some require a fairly high degree of reading comprehension. Chapter 6 focuses on a type of problem that is likely to be new to you, the data sufficiency question. Although these questions might initially seem discouraging, keep in mind that the only intimidating aspect is that the way they ask you to reason is novel. Chapter 6 gives you a solid strategy that, if consistently applied in practice, is sure to earn you more points on test day.

In an ideal world, you've given yourself enough time between now and the day of the exam to work thoroughly through this entire book. As a general rule, the more preparation time you put in, the better you're going to perform on test day. If you find, however, that time is not on your side, this book is still structured to accommodate you. It is ordered by topic. So just skip the topics you have mastered, skim the topics you're relatively familiar with, and submerge yourself in the chapters that contain concepts you're not quite comfortable with. The subject explanations have likewise been prepared with the busy test taker in mind and are organized by bullet points. Note boxes appear throughout the text to help you carry out some of the solutions more quickly without the aid of a calculator. Consider incorporating some of these tips into your solutions. We also suggest you prepare flashcards with some notes and explanations as you go through the subject explanations. Flashcards allow for more "mobile" study for those of you on the run.

Minimize Test Day Angst: Replicate Actual Testing Conditions in Practice

Now is the best time to start simulating actual testing conditions. Because you are studying at your own pace and will take the exam in a limited and unbroken time frame, you should minimize the discrepancies between the studying and testing situations. Before you begin a practice section, get yourself some scratch paper and do all of your calculations on that rather than in the book. This will help brace you for test day when you'll have no option but to look back and forth between the computer screen and the page on which you're doing your work by hand. Similarly, when you're taking diagnostic or practice tests (both of which we certainly suggest you do before test day), try the following. Set yourself a timer, turn your phone off, and let your roommates know they must not disturb you for the next four hours. Do not revert to using a calculator, even in practice. Acclimating yourself to doing everything by hand is certainly going to minimize the terror that the "no technology on test day" rule might otherwise provoke. In fact, we advise you to start doing all of your daily calculations by hand. The more mathematical problems you process without the aid of a calculator, the better.

If you are interested in simulating the exam even more precisely, GMATPrep test preparation software is available at *www.mba.com*. You have the option of downloading the software directly onto your computer or receiving a CD-ROM version of the software when you register for the exam. Two full computer-adaptive practice tests are included as well as real-time scoring. You have no reason not to take advantage of free practice exams, especially when the rule of this test is "practice like you're gonna play."

SECTION 1.4: SAY HELLO TO THE GMAT

The GMAT is a 3½ hour exam. In fact, a pre-exam computer tutorial and breaks bring it to a full 4 hours. The test consists of 78 multiple-choice questions and 2 short-essay writing assessments. Formerly created by the ETS (the infamous originator of the SAT), the test is now designed by ACT, Inc. The GMAT is developed in cooperation with the Graduate Management Admission Council (GMAC). The function of the GMAC is the implementation of guidelines, procedures, and policies about the admissions process and the types of intelligence business schools are looking for in applicants. The ACT then turns these guidelines into the kinds of questions you will encounter.

Although this book concentrates on the math segment of the exam, we think it is only fair to let you know what you can expect to see in each of the sections. The test contains 4 sections. Although the analytical writing portions can appear in either order, the quantitative segment will always precede the verbal segment.

1. **Analytical Writing Assessment:** Analysis of an issue (30 minutes)
2. **Analytical Writing Assessment:** Analysis of an argument (30 minutes)
3. **Quantitative (Math) Segment:** 37 multiple-choice questions—$\frac{1}{3}$ or more will be data sufficiency questions and $\frac{2}{3}$ or less will be problem-solving questions (75 minutes)
4. **Verbal Segment:** 41 multiple-choice questions split fairly evenly among critical reasoning, reading comprehension, and sentence correction questions, although there might be more sentence correction questions (75 minutes)

The kinds of verbal questions you'll encounter will not be covered by this book. The difference between the problem-solving questions and the data sufficiency questions is that: problem-solving questions are the kind you've grown accustomed to encountering on standardized tests. Your task is simply to solve the problem you are presented with and choose the correct answer from the 5 choices you are given. Feel free to skim the table of contents of this book to get a sense of what kinds of concepts we're talking about. Percentages, exponents, coordinate geometry, and quadratic equations are all possible topics. Data sufficiency problems also test quantitative reasoning. However, they use a set of directions you're going to want to have memorized for test day. Chapter 6 will discuss this in detail. In brief, though, you will be given a question followed by two statements. Both, either, or neither statement will be useful in determining the answer to that question. Your task will be to determine whether either statement alone will sufficiently answer the question, whether both are necessary to answer the question, or whether enough information is not in either of them to come to a conclusion as regards the question.

Breaks and Experimental Questions

You will be offered an optional 5- or 10-minute break after both the analytical writing and quantitative sections, which we suggest you take. Four hours is, after all, quite a long time to ask your body to be still and attentive. So you can't say you weren't warned, you should know that about a quarter of the questions you will encounter on the exam are experimental questions—questions that have no bearing on your score but that the ACT has slipped in to test how well they will be received on future exams. Essentially, you are paying the ACT to be its guinea pig. Even more unfortunately, you won't know which questions are the experimental ones. Just assume that all questions count and treat each question as one that will influence your score.

Your Scores and What They Mean

After you've hit the Answer Confirm button for the last time, the computer will calculate your scores for everything except the analytical writing portions. We won't brief you on how the testing screen will appear. You can access the tutorial on the GMAT website. At this point, you will have the option either to see your score, which will make it official, or to cancel it and never know how you fared. Of course, after all the studying you'll have done, you likely won't need to opt out. The two writing sections will be read by actual humans. The official score report—including the writing scores—will be mailed to you within 10 days after the exam. As we will explain in the next section, the scores aren't calculated by a straightforward "number of questions you got right minus number of questions you got wrong" method. Some questions are worth more than others. Each question's value is contingent upon its level of difficulty. Therefore, answering a difficult question correctly is worth more than answering a less difficult question correctly. Your official score will include:

1. Your quantitative score (from 0 to 60)
2. Your verbal score (from 0 to 60)
3. Your inclusive score (from 200 to 800, in 10-point increments)
4. Your analytical writing score (from 1 to 6, in half-point increments)

In the official report you receive in the mail, you will also be given a percentile rank (from 0% to 99%). The percentile rank is a reflection of your performance relative to all GMAT test takers in the most recent three years. A percentile rank of 70%, for example, means you scored higher than 69% of the other test takers in this time period. It also means you scored lower than the other 30%. As soon as your essays have been read and scored by the GMAT readers, score reports will be forwarded to each of the business schools you designated when you registered for the exam. Your scores will be transmitted to up to 5 schools without charge; sending then to each additional school will cost you a fee. Although most schools will simply average reported scores if you take the test more than once, some schools have begun disregarding scores that are disproportionately lower than others. Some institutions promise to look at only your highest scores, although this practice is becoming less common as it obviously favors those who can afford to take the test multiple times.

Good Scores vs. Bad Scores

The question we inevitably get asked regarding these reports is "What is a good GMAT score?" The short answer is that a good score is the best score that you personally are capable of achieving—something you'll definitely have a better sense of after taking a few practice tests. The longer and equally frustrating answer is that whether a score is good or bad is pretty indeterminate. No one has yet established a number by which to assess whether someone has passed or failed the GMAT. Although there is no benchmark, average grades have tended to differ based on undergraduate major. For example, physics and computer science majors tend to score the highest. Most scores tend to fall between 390 and 620. The average GMAT score in recent years has been around 530 or 540. GMAC publishes annual score reports for the year prior if you're interested in the details. We suggest you don't get too caught up in other people's averages. Instead, check in with the schools you plan on applying to. Each institution is likely to have a different number in mind for what scores their ideal candidates receive.

Time- and Stress-Saving Tips

To spare yourself any additional frustration on test day, you must learn now that you will not be allowed to bring anything into the testing room except the paper or note boards handed to you by the administrator. Cell phones, calculators (both handheld and on a watch), pens, watch alarms, and so on are prohibited. If you have any questions about the test itself, you can access a "Help" function at the bottom of the screen. However, your time will continue to expire even while you're getting answers to these basic questions. You will have an untimed period prior to the start of the test in which you are given an interactive tutorial to allow you to become familiar with the program you are about to use. This tutorial includes information about how to use the word processing program that you'll use to write your analyses. Presumably, you'll have all the time you need during this tutorial to go over the details. If you're going to be sitting there for 4 hours already, there isn't much of a point in spending an additional half hour to learn the system. Free test tutorials that will acquaint you with the basics, such as entering answers, are available at *www.mba.com*. Getting acquainted with the procedures prior to the test is going to be critical—and is going to reduce your stress tremendously on the big day.

You might want to familarize yourself with the other testing procedures and rules before arriving at the test site. They include fingerprinting, identification, disclosure prohibition, and so forth. The list is long. If you're interested in the details, you can access all of them at *http://www.mba.com/mba/thegmat/testday*.

SECTION 1.5: HOW THE CAT WORKS

CAT is an acronym for computer-adaptive test. Although the computer and the test aspects of the name should be pretty straightforward, you should be most concerned with the adaptive aspect. Adaptive essentially means that there is no predetermined order of difficulty to the questions you will be given. This is the biggest difference between the CAT and a paper exam. The computer tries to match the difficulty of each successive question to both your overall performance and to the question you've just answered. The computer takes into account not only whether or not you answered the previous question correctly but also the difficulty level of that particular question. Thus the better you perform, the more challenging the questions will get—and the more challenging the questions, the more points you will receive for answering them correctly. In short, your score is not determined by the number of questions you answer correctly but by the difficulty of the questions you answer correctly.

The first question you will see is one that the computer considers of average difficulty—a question 50% of test takers are likely to answer correctly and 50% are likely to answer incorrectly. If you answer that question correctly, the computer is going to give you a slightly more difficult question the next time around. If you answer it incorrectly, though, your second question will be slightly easier.

Although the computer assumes from the beginning that you are an average examinee, it determines pretty quickly whether or not this is actually the case. For each subsequent question, the program will continue to choose from a large pool of test questions, all arranged by question type and difficulty. The particular question chosen depends on how you seem to be handling the given questions. The more questions you answer, the better refined the computer's interpretation will be of the difficulty level you are capable of negotiating. By the time you're halfway through the quantitative and verbal sections, the computer should be choosing questions that are in accordance with your aptitude. This means that most of the questions will seem challenging and you will get approximately as many correct answers as incorrect answers. This will be the level the computer uses to determine your scaled score.

What the Structure of the CAT Means for You as a Test Taker

You must understand a few crucial things about how this format differs from the kind you're used to taking. Most importantly, you can't skip questions on the CAT. You will be able to see only one question at a time. You cannot go back to change an answer once you've confirmed it. Because the computer predicates each succeeding question on your answer to the question prior, you have no choice but to answer the question in front of you—even if it means guessing—and let it go. (We will talk later about strategic guessing so you can choose intelligently when you don't exactly know the answer.) Do not get frustrated if the test seems to be getting incrementally more challenging. If it does, this means that you're doing well. The last thing, though, that you want to do on test day is try to ascertain whether the question you've got before you is easier or harder than the last one. Trying to figure out how the computer thinks you are doing is only going to take you away from your real focus, which is performing your best in the moment and on the present question.

Giving the Early Questions More Time and Energy Than the Later Ones

Probably the most critical thing you need to keep in mind is that your final score is by no means based solely on the last question that you answer. In fact, given what you now know about the algorithm the test uses to determine each question, it is more important to answer questions correctly early on than to answer questions correctly at the end. In fact, we highly suggest that when you're allocating time, you spend a disproportionately larger amount of time on the first 10 questions than on subsequent ones. You want to aim for 90% to 100% accuracy, in particular, on these first 10 questions. Why is answering the first 5 questions correctly going to be more advantageous than answering the final 5 questions correctly? Remember, the more difficult questions are worth more points. So if you start generating the difficult questions at the beginning, you'll be getting more points for each question you answer correctly both in the middle and at the end of the exam. If you don't start generating difficult questions until the middle of the exam, you'll be receiving additional points only for those questions you answer correctly at the end. Simply put, the harder questions are where the points are. The more hard questions you encounter and the sooner you encounter them, the higher your score will be. Also keep in mind that this does not mean that if you miss the first question, you're done for. The more questions you tend to answer correctly, the more likely the computer is to recognize any silly mistakes you make as anomalies. So do not fear a one step forward, two steps back progression.

Leave No Question Unanswered

Please keep in mind that you will be penalized for all questions that you leave unanswered on the CAT. In fact, each question you fail to answer will decrease your score by an even greater increment than each question you answer incorrectly. This means that if you're running out of time on a section, you're going to earn more points by merely guessing than by letting time run out. We know that conceptually this is a little difficult to accept. Why answer a question you've hardly glanced at? The GMAT is likely one of the few cases in which this is ever going to be true. However, random guessing is, in this case, better than nothing.

Our Final Word on the CAT

As you've likely gleaned from this introduction to the CAT, it's got a few pros and a few cons relative to paper-based tests. Among the cons of the CAT are that you can neither skip nor return to a question. Additionally, you can't make notes on the computer screen. You can't see all of the reading comprehension passage or the charts at once; you've got to scroll. A compelling aspect of the medium is that test takers who have more experience using computer programs will have a bit of an advantage. However, a fair number of pros essentially tip the balance back. Because it is a computer test, you can take the CAT at a time that is convenient for you. You also learn your verbal and quantitative scores almost immediately after you've answered the final question. You have a timer on the screen for purposes of pacing and have a more personalized testing space (a computer alcove). Preparing yourself to overcome the few cons, we think, will be worth your while.

SECTION 1.6: PREPARATION STRATEGIES

- **First things first: take a diagnostic test.** You will not be able to set your goals or be able to organize a study schedule for yourself without having a sound idea of what level you are at the start. There's no reason to spend all your preparation time studying grammar if you already know everything you're going to need to know. A worse scenario is finding out that much of what you studied is not even material the GMAT covers. Conversely, you might find out through a diagnostic test that you don't remember anything about parallel lines and transversals. Setting aside the time to take a full diagnostic test will help you establish exactly where you should expend your energy over the coming weeks. It will also give you a sense from the outset of the structure and the language you can expect on the exam.

- **Sign up and set your goal.** Another crucial step we strongly suggest you take before organizing a study plan is to decide on a convenient and suitable time to take the test and actually register for it. Determining the timing of the GMAT is somewhat of a chicken and egg problem where you will need to consider various factors. These include your work schedule, how much time you estimate to spend working toward your goal and the timing of the test. So make some reasonable estimates on your time commitment and register in advance. This will not only ensure that you take the test at your convenience but will help you determine the details of your study plan.

    Thinking about these factors together with the general score expectations of your favorite schools will also help you set specific, achievable, and realistic goals for your score.

- **Create a realistic study plan for yourself and stick to it.** Only you know how well you did on the diagnostic test, and only you know how much time you have until test day. Get yourself a calendar and put it on the fridge, above your desk, or in any conspicuous place. Keep in mind that you have a life to live and other responsibilities. Schedule an hour a day or a couple of hours two days a week to study. You can even devote certain weekends to certain concepts if necessary. Having this schedule is going to be useful in two ways. First, it will ensure that GMAT study doesn't get last priority in your life. We all know how tragic that "I'll just finish these other things first" mentality can be. Second, looking at the big picture will help you realize that, in fact, you have enough time to prepare. Seeing your study schedule stretched out over a period of time will keep it from seeming so daunting.

- **Consider your shortcomings.** Remember that your GMAT scores are going to the institutions you are applying to in their entirety. You can't simply transmit a section of the exam. We know how easy it is to say "I understand 75% of this stuff; I'll brush up on the tougher material a few days before the exam." We also know Murphy's law states that the first questions you'll get on exam day will cover precisely the areas you waited until the last minute to study. If you use this book, the GMAT website, and all the other sources of information out there, there's no way you can't know what is going to be on this exam. This should be satisfying news. So once you've taken the diagnostic test, focus first on the question types that gave you the most difficulty. These questions made you pause the longest before answering. You had to read them twice before you understood them.

- **Start doing all your calculations by hand today.** We've already advocated this minor shift in your daily life, so we won't harp. Practicing your multiplication tables rather than relying on the calculator in the coming weeks is going to be the equivalent of many precious GMAT minutes.

- **Whenever you take a practice test, try to reproduce an authentic situation as closely as possible.** Use a word processer, if you have access to one, and use only the features that will be available to you on the CAT. (The analytical writing section will be typed into a similar program.) It will be good to have a sense from the get-go of how your body responds to sitting in a chair and keeping your focus intact for 4 hours straight. This is definitely not the same thing as watching 4 consecutive hours of television. You really need to determine, for example, whether your pacing is where it should be. Also remember that any computer-based exams you can find are going to provide you with more accurate conditions. So if you have a choice between practicing on the computer and practicing in a test booklet, always choose the former. If you are computer challenged, this is going to be crucial. Borrow a computer from a friend if you need to get used to using a keyboard or mouse. Being comfortable with the medium is key.

- **Think of the GMAT as a simultaneous ultramarathon and 200-meter sprint.** You're going to have to answer each question in about 2 minutes. We know that if you had all the time in the world to complete the GMAT, you'd likely score in the 99th percentile. Unfortunately, the GMAT writers know that too, which is why they've introduced the time limitation. Maintaining an aggressive and industrious pace is one of the few things that will distinguish you from your fellow test takers. At the same time, the GMAT requires a level of mental stamina that you will likely need even more of once you get to grad school. If you can strike a balance between these two modes, you've already outdistanced a good number of GMAT examinees.

- **Know the directions well enough that you can recite them in your sleep.** This is paramount, especially for data sufficiency questions. The CAT will display the directions for each question type as it appears on the screen. However, the last thing you want to waste your time on is reading the directions. If you've already familiarized yourself with what each section is going to require of you, there will be no need to let the clock keep running while you make certain you're doing the work the question actually entails.

- **Now is not the time to make any (other) life-changing decisions.** Don't try to quit or start smoking, don't move to a new apartment, don't begin training for a triathlon, and don't go through a breakup in the weeks before the exam. If at all possible, sign up for the GMAT at a time you know you won't have too many other responsibilities or priorities to juggle. This exam is going to take quite a lot of mental stamina and emotional stability. Ensure that you're burdened with as few things as possible during your preparation.

- **Take the GMAT early enough that if you need to retake it for any reason, you have time to do so.** Most MBA programs admit students for only fall matriculation. November is generally the latest month you can take the GMAT if you want to ensure your scores make it to all your schools in time. You might want to think about taking it earlier than November, however, in case something goes awry with your November test. For instance, you may come down with the flu, you may not do as well as you expected, a family emergency may take you out of town, and so forth. The GMAT process is the same as any other life process with deadlines. Don't wait until the last minute or the last sitting. Always keep your "in case" mentality active.

- **Know how to minimize anxiety.** Are you worried about timing? Complete a few more practice tests under exam conditions. Are you concerned about how your scheduling and study habits compare with those of others? Join a GMAT study group. Are you agonizing over whether you've given yourself enough time to prepare? Talk to people who have taken similar exams. They're out there, everywhere. You'll find your insecurities will dissolve pretty quickly. Have you been studying for such long hours that you're starting to get "easy" questions wrong? Get on your bike. Quiet your thoughts. Step away from the desk for awhile. If you need to keep the GMAT in the forefront of your mind, go out for coffee with a friend and correct his or her grammar. Instead, you can choose one of the possible "Analysis of an Issue" topics and answer it in your head while you're taking a walk through the park. Whatever you need to do, remember that you're still in the world and that the GMAT is not the be-all, end-all. Your mind is also not structured to handle 12 hours of straight study. Break it up a bit. All of the above will help you maintain a positive attitude, which is really the most important aspect of this process.

- **Keep your scores and your expectations in perspective.** If you're picking up this book because you've already signed up for the GMAT or are about to, you've likely also set your sights on a few colleges or universities you'd love to study at. You probably also have a good idea what kind of scores those institutions require. If you haven't looked at those schools' requirements, you should. Doing so will give you a sense of where you are relative to the schools' general expectations. Knowing what your target schools consider acceptable scores is a good idea. However, you shouldn't obsess over those numbers. Don't sabotage yourself with the numbers the school provides for you as a gauge. Looking at the scores of your diagnostic test is much more constructive. With each new test you take, consider how you can raise your score a bit from last time. Think of studying as a competition with yourself. There's not too much value in putting yourself against the rest of the GMAT test takers worldwide. Setting realistic expectations, realizing what your limitations are, and so forth are going to be essential to ensuring you neither disappoint yourself nor work yourself into a frenzy.

- **Know when you've reached your optimum score.** A time will come when you've taken a number of practice tests in a row and you just keep getting the same score, despite the amount of studying you're doing. Ideally, this will be around the same time your test is scheduled. Don't drag out the process by continuing to postpone the test so you can get in more preparation. At some point, your motivation, your interest, and your performance will start declining. (So might other things, like your social skills.) You've brushed up on the math and the grammar. You've rectified any bad test-taking habits. You've learned how to pinpoint wrong answer choices instinctually. You also know you can finish the exam in a timely manner. What else is there? Take the test—for your own sake—before you start driving yourself crazy.
- **The day before the test, congratulate yourself instead of cramming.** We all know from past experience how unproductive burning the midnight oil actually is. If you've given yourself the time to prepare, stuck to your schedule, and eaten, slept, and continued to breathe throughout the process, you're in great shape. Take yourself out to your favorite restaurant the evening before the GMAT or go with a friend who has supported you during your preparation time. Eat the kinds of foods that make you feel good in the morning. Relax and get to bed early. Remember the ultramarathon part of the test. The exam won't slow down for you if you haven't slept the night before because you've convinced yourself there are a few more things you have time to insert into your short-term memory.

SECTION 1.7: TEST DAY STRATEGIES

- **Stick to your usual routine on the day of the test.** If you typically have oatmeal and juice for breakfast, eat that. On the morning of the GMAT, don't have pancakes and coffee if that's not a usual breakfast for you. Also be sure to time your pre-GMAT meal. Don't eat right before the test because you don't want all your energy going to digest the food in your stomach when it should be circulating blood to your brain. You should eat about an hour or so before the exam begins. Definitely bring a snack if you think you'll need a midtest pick-me-up. You won't be allowed food at your alcove, but you can bring it to the testing center to eat during one of your two short breaks.
- **Devote special attention to the first 5 to 10 quantitative and verbal questions.** The details behind this strategy were previously described. Briefly, though, the CAT tends to discover within the first questions of each section what difficulty level you are capable of maintaining. The higher the CAT believes your competency level to be—and the sooner—the higher your final score is going to be.
- **Use your scratch paper, and use the tutorial time to prepare it.** One of the most frustrating aspects of the GMAT is that you spend most of the test looking back and forth between the computer screen and your scratch paper. This should not mean you don't make use of the paper. Doing work in your head almost always spells disaster in some form or another. Before the test starts, perhaps during the untimed interactive tutorial at the beginning, divide your scratch paper into a chart to help you use the processes of elimination. Obviously, you can't cross out answer options on the computer screen. So the best way to combat this is to divide your paper into a chart with 5 sections. In each box to the left, write A, B, C, D, and E so that you can physically cross out the answers you decide to eliminate (leave a blank space at the bottom for formulas and such). Using a paper chart is great way to keep track of the remaining feasible answer choices

as you work out the problem. It is much better than covering the answers you decide to eliminate by placing your fingers onto the screen.

- **Process of elimination is your new best friend.** We mean this. The sooner you can get rid of the answer options that are definitely not right, the sooner you've got a significantly higher chance of choosing the right answer. Frankly, determining which answers are wrong is often much easier than determining which answer is right. Some of you might remember the satisfaction of reading a question on some previous standardized test and crossing out three of the answers immediately. Few things are more gratifying than knowing you've just increased your 20% chance of choosing the correct answer to a 50% chance. So don't be afraid to come at the answers in this backward fashion. Aggressive elimination is critical to performing well on a timed test such as the GMAT. We highly suggest this method if for no other reason than that it gets the wrong answers out of the way—and out of your mind.

- **Remember the verbal strategies.** We'll give you a few examples even though this is a math workbook. Note that specific math tips are in Section 1.8. First, on sentence correction questions, you can skip the first answer choice. The first answer option is always going to repeat the original. There's no sense reading it a second time if you know the original is incorrect. Second, read passages for the general outline, not for the details. Most of the questions are concerned with the passage structure, not its minutiae. Keep this in mind for these sections. Third, in the critical reasoning questions, find the conclusion first. The GMAT loves main point and primary purpose questions. The conclusion will be the key to answering every question of this type correctly. Finally, have two solid examples for each of your analytical essays. Every possible prompt you will respond to can be accessed on the GMAT website at *www.mba.com*. Take a look at some of these to get a sense of the kind of questions you will be asked.

- **Maintain an active mental process.** It becomes a little too easy—especially an hour or two into staring into the light of the screen—to let your eyes passively rove. Maintaining an attentive and inquisitive mind throughout should help with this. Ask yourself the following when you encounter each new question.

  - In what area is the computer trying to measure my proficiency by asking me this question?
  - What skills and intelligence am I being asked to apply right now?
  - What are the data, and what is the unknown?
  - How would a less careful test taker be flustered by a question like this?
  - Which of these answers are meant to mislead careless testers?
  - What is the most direct route I can use to find the answer to this question?

- **Read each question and each answer choice thoroughly.** We can't even begin to tell you how many test takers we've watched answer questions incorrectly simply because they read answer options A, B, and C but not D and E. Choice C sounded right, so they chose it without looking at D and E. Unfortunately, choice D was the correct answer. In fact, not reading everything fully is one of the leading causes of incorrect answers on the exam. Remember, as "easy" as the multiple-choice format might appear, all the answer options are written by the test makers because they seem like they could be correct. Every answer is going to seem like a possibility. Don't get caught up in an enticing answer choice and disregard the others before you give them a chance. Likewise, be sure you know what the whole question is asking. Many of the quantitative problems, in particular, have to be answered in two parts. If you answer only the first part, even if you're doing the work right, your answer is going to be wrong. We guarantee that one of the answer options is going to be that answer you arrive at when you're only halfway there. It's one of the tricks the GMAT writers love to play on you.

- **Move the keyboard out of your way once you are finished with the analytical writing section.** It will no longer be useful to you. In fact, those unwieldy things are likely to get in the way. The only two things you are going to need to focus on for the remainder of the exam are the computer screen and your scratch paper. Keeping your alcove otherwise clutter free will simultaneously de-clutter your mind.

- **Pay attention to your timing and your pacing throughout the test.** Remember the test is designed in such a way that you'll always be working a little faster than is comfortable for you. Ideally, you'll have taken enough practice tests that you have a solid sense of how you keep up. Also remember that you've budgeted a little more time for those first 10 questions and that you're going to have to pick up the pace significantly as you get into the second half of each section. You absolutely should spend more than average time on each of these initial questions as long as you have worked out how to make up that time. Do not let yourself spend too much time on any one question in particular. At a certain point, you're going to want to make your best guess and move on. Do not dwell too much on the clock. However, glance at it after every 5 questions or so. You'll be able to adjust appropriately from there.

- **Learn the art of intelligent guessing.** Inevitably, you are going to have to guess sometimes. Remember, the test is structured to give you precisely the questions that are difficult for you. It may be a matter of not knowing absolutely what the question is asking, having forgotten a formula, or simply running out of time. The more practice tests and questions you do, the better you will get at maximizing your chances of guessing correctly. The process of elimination, estimating, and recognizing the answers that the ACT anticipates average examinees will choose are all methods by which to make prudent guesses. We'll cover some of these in the following section.

- **Guess wisely and move on.** Remember two things. First, the next question you answer correctly is going to bump you back up to a harder question and more points. Second, more than 20% of the questions in both the verbal and quantitative sections are experimental questions. You may feel better if you keep in mind there is a good chance that any question you've had to guess on isn't ultimately going to affect your score at all.

- **Make use of the breaks—but remember the clock is always running.** You might finish a section of the test, determine that you are on a fabulous roll, and think you have enough momentum to rush headlong into the next section. Remind yourself in these moments that that next section is over an hour long. An unbroken 4 hours of anything is a lot. We recommend stretching to get your circulation flowing. Remember, your mind won't concentrate if no blood is circulating to your brain. Grab a snack or focus on some deep breath work. Don't use the breaks to mourn over questions you think you missed; let go of those. Congratulate yourself on the portions of the exam you've completed so far. Know that the minutes go by quickly and the test is going to resume again after the break is over—whether or not you are there.

- **Be careful about going anywhere near the "Section Exit" and "Test Quit" commands.** If you confirm either, the test is as good as canceled. The only buttons you'll really need will be the "Answer Confirm" and "Next" buttons, the ones on the right side of the screen.

- **If for some reason the test does not go as planned, you will have the option to cancel your scores at the end of the exam.** You will never know what your scores would have been.

However, if you are certain these aren't the scores you want going out, this is the price you'll have to pay. Before you hit the "Cancel" button, remember that the computer is structured to give you questions that are difficult enough that you're bound to miss some. The more trying questions mean you're on your way to a high score. Many GMAT takers think their performance was poorer than it actually was for this reason. Chances are high that if you've put in the work, you're going to be pleasantly surprised if you decide against score cancellation.

- **Plenty of staff will be at the testing site should something go wrong.** These problems include computer malfunctions, distractions, and so forth. Don't worry, though. The designated centers are generally well run, so the risk of any of these is low. Know that you can notify the staff with any questions or complaints and that reports can be filed online to ACT. So you're not without recourse in the case of unanticipated complications.

- **Congratulate yourself on the way out (and all that evening, and all the next day. . .).** You have just completed an exam that only a crazy few have the mental and emotional stamina to take. Remember to thank all the people that have constituted your support system in the past weeks and months. Continue to remind yourself in the coming weeks when you're waiting for your official score what a truly impressive feat you've just accomplished.

SECTION 1.8: SPECIFIC TIPS FOR THE QUANTITATIVE SECTION

- **Be particularly attentive to the two-part math questions.** It's easy to forget that the quantitative section actually takes a fair amount of critical reading. Peruse these questions as carefully as you do those in the verbal section. Here is an example of a problem in which a simple reading error could get you in a lot of trouble:

If $3x - 17 = 151$, then $x + 4 =$

(A) $48\frac{2}{3}$

(B) 52

(C) 56

(D) 60

(E) 172

So you start solving the equation, right? Add 17 to 151 to get $3x = 168$. Divide both sides by 3 to get $x = 56$. Bingo! We know what x equals. Furthermore, and this is where GMAT will get tricky with its answers, the number 56 is one of the choices. Option (C) is obviously (?) the answer.

It's really amazing how regularly this happens, especially when test day jitters are combined with serious time limitations. An attentive test taker will reread the question after finding the answer to ensure he or she found the answer to the question actually asked. Reread the question. It is not asking what x is but, rather, what $x + 4$ is. The correct answer is (D).

Notice the other answers are all answers you could potentially get if you weren't attentive in your calculating. This is another reason why you must use your scratch paper. You might have chosen (E) if you had added 17 to 151 to get 168, but had forgotten to divide by 3. You would have determined that $x = 168$. You would instead have chosen (B) if you made it all the way to $x = 56$ but then subtracted 4 instead of adding it. You would have arrived at option (A) if you had subtracted 17 from 151 rather than added it. ACT expects that a certain number of test takers are going to make mistakes of this sort on the math section. Be sure you are working carefully but quickly. Don't find yourself trapped into thinking a particular answer looks good because you incorrectly arrived at the exact same answer.

- **Don't be afraid to work backward from the answers on quantitative problem-solving questions.** Because your answer choices are all right there, the correct answer is necessarily staring you in the face. So make use of your options! Plug in each answer one at a time; one of them is bound to fit. One thing you might have noticed, too, is that the answer choices are always arranged in size order. Starting with choice (C) will often allow you to cross out the two choice above or the two choices below and then move on. Here's an example:

Kumi is 4 times as old as Aimee. 3 years ago, Kumi was 5 times as old as Aimee. How old is Aimee today?

(A) 9
(B) 12
(C) 15
(D) 45
(E) 48

This is the perfect example of a question that you could plug in answers. Who wants to start pulling numbers out of thin air to see if they work? Let's start with (C) and assume that Aimee is 15 years old today. Remember, that is what the question is asking. The first sentence says that today Kumi is 4 times as old as Aimee. That would mean that Kumi is 60 years old. Our next sentence tells us that 3 years ago, Kumi was 5 times as old as Aimee. So 3 years ago, Kumi was 57 (60 − 3) and Aimee was 12 (15 − 3). Is someone who is 57 years old 5 times as old as someone who is 12? No. So we can cross out (C) and move on.

Now, if we're thinking clearly, we know that the numbers we arrived at for (C) are quite close to where we want to be: 12 times 5 is 60, and 57 is very close to 60. So we're going to want to stay in the range of 15. In fact, we can cross out answers (D) and (E) because they are nowhere close. This is called aggressive elimination, something that is going to be your best friend for this exam. We could also consider that if Aimee is either 45 or 48, Kumi would be close to 200 years old. The GMAT isn't going to present you with impossible circumstances. Let's instead work backward and try (B). If Aimee is 12 years old today, then Kumi is 48 years old. So 3 years ago, Kumi was 45 (48 − 3) and Aimee was 9 (12 − 3). Is someone who is 45 years old 5 times as old as someone who is 9? Absolutely. We know our answer is (B), and we really had to consider only two answer options.

Remember that working backward from the answer choices may take considerably longer than using an algebraic solution, especially when you cannot eliminate any of the answer choices from the beginning. Work backward sparingly. Use the process of elimination first where possible.

- **Plug in numbers if the question gives you only letters.** These are the fun ones, relatively speaking. You get to assert some kind of control by choosing the numbers you want to use. Consider this method as an option whenever you are presented with a question that is asking you about the relationship between any set of letters. Here's an example:

If $x + y = 24$, and $x − y + z = 15$, then $4x + 2z =$

(A) 6
(B) 9
(C) 45
(D) 78
(E) 126

That's a lot of letters and not a whole lot of numbers. These questions can be radically disconcerting until you realize that you get to choose the numbers that you want to work with. Essentially the problem is in your hands. This is also where choosing numbers wisely comes in. You don't want to give yourself any more work than is necessary to complete the problem.

So we've got two equations with three unknowns since nobody has given us a single clue as to what x, y, and z are. The only thing to do is to start plugging in your own numbers while staying within the constraints you are given. The first equation we've got is $x + y = 24$. This means our only constraint is that the two numbers we pick for x and y have to add up to 24. Now here's what we mean by choosing wisely. A lot of potential number combinations will get us to 24. However, it makes no sense to choose, say, 114 and -90. These numbers might only complicate things for us further on. So let's let $x = 20$ and let $y = 4$.

We've let $x = 20$ and $y = 4$, and we know that $x - y + z = 15$. We can now plug in the numbers we've determined for the first equation into the second: $x - y + z = 15$ can now be rewritten as $20 - 4 + z = 15$. Subtract the numbers on the left. We're down to $16 + z - 15$, which is a much easier equation to solve. Now we know that $z = -1$.

The hardest part is now complete. We've determined that one of the possible combinations of x, y, and z are $x = 20$, $y = 4$, and $z = -1$. Now fill in the expression $4x + 2z$. It can be rewritten as $4(20) + -2$, which can be solved. The answer is 78.

Alternatively, you can always use algebra to determine $4x + 2z$. To answer the question, you do not need to find x and z separately but just need to find $4x + 2z$ directly. Add the first two equations side by side to eliminate the y-variable:

$$
\begin{array}{r}
x + y = 24 \\
+\ \underline{x - y + z = 15} \\
2x + z = 39
\end{array}
$$

Since the question asks for $4x + 2z$, multiply both sides of the equation by 2 to get $4x + 2z = 78$.

The more math practice you do, the more you'll enjoy questions made up almost entirely of letters. You need to recognize the relationship among the set of letters the GMAT is giving you, pay attention to the constraints, and choose numbers that are going to make your calculations as easy as possible.

- **Remember that estimating can often eliminate a few answers for you right away.** The GMAT doesn't care how you get to the correct answer. All that it registers is whether or not you got there. This means that estimating is not the equivalent of lazy math. Rather, it is a smart and quick way of eliminating those excess (wrong) answers almost immediately. Estimating is sure to save you loads of time. Let's take a look at this question:

A mixture of red beans and black beans is to be prepared. The price of red beans is $2 per pound, and the price of black beans is $3 per pound. What is the ratio of red beans to black beans if the mixture is to be sold for $2.75 per pound?

(A) 1:3

(B) 1:2

(C) 2:3

(D) 1:1

(E) 3:1

Applying a little common sense before we really start messing the scratch paper with unnecessary calculations will come in handy here. The price of the mixture is $2.75, which is closer to the price of black beans ($3) than the price of the red beans ($2). That means the mixture has more black beans than red beans. Since the question is asking us the ratio of red beans to black beans, our answer needs to be a fraction that is less than 1. Thus we can get rid of any number that is equal to or greater than 1. Cross off (D) and (E). See how quickly we're down to just 3 answer choices.

All we really mean by estimating is recognizing that your answer has to fall into a particular range of possible numbers. Once you come to that conclusion, everything that falls outside of that set is suddenly no longer an option.

- **Always look at the answer choices before you begin on the question.** The answer choices will give you a sense of how much work you need to do. For example, if all of your answer choices contain π, you're not going to want to spend your precious time converting everything to 3.14 only to have to undo it again when it's time to choose your answer.

- **Know the data sufficiency answer choices cold before you walk into the exam.** Ultimately, you're going to be able to approach each data sufficiency question in the same way. We suggest you pay particular attention to Chapter 6 if you've never seen a question of its kind before, which is likely if this is your first go at the GMAT.

SECTION 1.9: REGISTERING FOR THE GMAT

You can take the computer-adaptive test 10 hours a day, 6 days a week, 3 weeks a month at over 400 testing centers worldwide. You shouldn't have too many scheduling conflicts with all of those options. You can register for the GMAT online at *www.mba.com* if you've got a credit card handy. You can also call one of the many test centers listed online at *www.mba.com* if you prefer to register in real time with an individual. You may also call 1-800-717-GMAT if you're in the U.S. or Canada. The registration fee as of 2010 is $250. Because of the nature of the CAT, it's certainly possible to register for an exam even up to a few days before. Remember that test centers start filling up in October. Finding a convenient center becomes increasingly difficult in November as MBA applicants are finalizing their entrance requirements. Either way, we suggest you schedule a couple of months or at least a few weeks in advance. You want to ensure, after all, that your test scores reach the institutions you're applying to in time. Knowing how much time you have before taking the GMAT is going to be crucial when working out your study schedule. It also gives you a definitive deadline by which to consider yourself prepared. Lastly, consider your biological rhythms when deciding on the time of day that's best for you. If you're a morning person, schedule an early appointment. If you don't start waking up until noon, keep that in mind. Arriving at the test center at the time of day when you are at your peak is going to be the final strategy in a long line of pivotal decisions. Follow through on all these decisions while on the road to acceptance into business school.

Arithmetic

2

SECTION 2.1: PROPERTIES OF REAL NUMBERS AND INTEGERS
Real Numbers

- Real numbers include all numbers on the number line ($0.2, 4, 333, \pi, \sqrt{3}, -\sqrt{5}, \ldots$).

- Rational numbers are all numbers that can be represented as a fraction $\left(\frac{2}{3}, 0.004, -12.3, \frac{66}{05}, \ldots\right)$.
- Irrational numbers are all numbers that are not rational ($\sqrt[3]{2}, -\sqrt{3}, \sqrt{5}, \pi, \ldots$). They cannot be represented as fractions.
- Repeating decimals are rational numbers $\left(0.\bar{3} = \frac{1}{3}, 0.\overline{66} = \frac{2}{3}\right)$.
- Integers are all numbers that have no decimal or fractional components ($-3, 878, -99, \ldots$).

  Note that every integer is also a rational number. Integers are a subset of rational numbers.

ABSOLUTE VALUE

- The absolute value of any nonzero number is always positive. The absolute value of x is written as $|x|$.

  Example: $|-4| = 4$, $|17| = 17$, $|0| = 0$

- For any number, $|-x| = |x|$.
- If $|x| = 7$, then $x = 7$ or $x = -7$.

> **REMEMBER**
>
> If x is positive, then $|x| = x$.
> If x is negative, then $|x| = -x$.
> For example,
> if $x = 5$, then $|5| = 5$ and
> If $x = -3$, then $|-3| = -(-3) = 3$.

PROPERTIES OF REAL NUMBERS

- **Commutative Property:**

  $a + b = b + a$ and $a \cdot b = b \cdot a$

  $3 + 12 = 12 + 3$ $17 \cdot 5 = 5 \cdot 17$

- **Associative Property:**

  $a + (b + c) = (a + b) + c$ and $a \cdot (b \cdot c) = (a \cdot b) \cdot c$

  $13 + (12 + 14) = (13 + 12) + 14$ $22 \cdot (2 \cdot 6) = (22 \cdot 2) \cdot 6$

  $13 + 26 = 25 + 14$ $22 \cdot 12 = 44 \cdot 6$

  $39 = 39$ $264 = 264$

- Use commutative and associative properties to simplify certain calculations.
  First add the numbers that add up to 10 or a multiple of 10.

  Example: In $(17 + 8 + 13 + 12)$, rearrange to perform easier additions:

  $(17 + 13) + (8 + 12) = 30 + 20 = 50$.

  Multiply even numbers and the number 5 first wherever possible $(4 \cdot 5 = 20, 2 \cdot 5 = 10)$.

  Example: For $2 \cdot 3 \cdot 15$, multiply 2 and 15 first.

  $2 \cdot 3 \cdot 15 = 2 \cdot 15 \cdot 3 = 30 \cdot 3 = 90$

- **Distributive Property:**

  $a \cdot (b + c) = a \cdot b + a \cdot c = ab + ac$

  Example: $3 \cdot (x - 5) = 3 \cdot x - 3 \cdot 5 = 3x - 15$

- The distributive property is also very useful when simplifying certain calculations.

  Example: $12 \cdot 19 = ?$

  Instead of trying to multiply by 19, write 19 as $(20 - 1)$.

  $12 \cdot 19 = 12 \cdot (20 - 1) = 12 \cdot 20 - 12 \cdot 1 = 240 - 12 = 228$

  Example: $63 \cdot 11 = 63 \cdot (10 + 1) = 630 + 63 = 693$

Integers

Numbers that are included in the set $\{\ldots, -4, -3, -2, -1, 0, 1, 2, 3, 4, \ldots\}$ are called integers.

- $\{\ldots, -4, -3, -2, -1\}$ are negative integers.
- $\{1, 2, 3, 4, \ldots\}$ are positive integers.
- The integer 0 is neither positive nor negative.

> **REMEMBER**
>
> The integer 0 is neither positive nor negative. It is even.

ODD AND EVEN INTEGERS

- Integers divisible by 2 are even integers: $\{\ldots, -4, -2, 0, 2, 4, 6, \ldots\}$
- Integers not divisible by 2 are odd integers: $\{\ldots, -5, -3, -1, 1, 3, 5, \ldots\}$

Multiplication		Addition and Subtraction	
(even) $\cdot$ (even) = (even)	$-8 \cdot 2 = -16$	(even) $+/-$ (even) = (even)	$4 + 12 = 16$
(even) $\cdot$ (odd) = (even)	$4 \cdot 7 = 28$	(even) $+/-$ (odd) = (odd)	$-12 + 9 = -3$
(odd) $\cdot$ (odd) = (odd)	$11 \cdot -3 = -33$	(odd) $+/-$ (odd) = (even)	$5 + 17 = 22$

- There are no even/odd rules for division. One number may not be divisible by another number. Also, an even number divided by an even number could be either even or odd.

    Examples: $12 \div 2 = 6$ (even $\div$ even = even)
                    $6 \div 2 = 3$ (even $\div$ even = odd)

CONSECUTIVE INTEGERS

- Consecutive integers are integers that $\{-3, -2, -1, 0, 1\}$ or $\{61, 62, 63\}$
  follow in a sequence:
- Consecutive even integers are even integers $\{-12, -10, -8\}$ or $\{44, 46, 48, 50\}$
  that follow in a sequence:
- Consecutive odd integers are odd integers $\{-11, -9, -7\}$ or $\{147, 149, 151\}$
  that follow in a sequence:

When solving word problems, consecutive integers can be symbolized by $n, n + 1, n + 2, \ldots$.
Consecutive odd or even integers can be symbolized by $n, n + 2, n + 4, n + 6, \ldots$.

PRIME NUMBERS

- A prime number is an integer greater
  than 1 and divisible by only 1 and itself: $\{2, 3, 5, 7, 11, \ldots\}$
- 1 is not a prime number.
- The smallest prime number is 2.
- 2 is the only even prime number.

MULTIPLICATION AND DIVISION OF SIGNED NUMBERS

(positive) $\cdot$ (positive) = (positive) $8 \cdot 2 = 16$
(negative) $\cdot$ (positive) $=$ (negative) $-4 \cdot 7 = -28$
(negative) $\cdot$ (negative) = (positive) $-11 \cdot -3 = 33$
(positive) $\cdot$ (negative) = (negative) $18 \cdot -2 = -36$

(positive) $\div$ (positive) = (positive) $8 \div 2 = 4$
(negative) $\div$ (positive) = (negative) $-4 \div 2 = -2$
(negative) $\div$ (negative) = (positive) $-4 \div -2 = 2$
(positive) $\div$ (negative) = (negative) $4 \div -2 = -2$

ADDITION OF SIGNED NUMBERS

- If two numbers have the same sign, add the absolute value of the numbers and keep the sign.

    Examples: $(-5) + (-7) = -12$
                   $42 + 28 = 70$

- If two numbers have different signs, subtract the absolute value of the numbers and keep the sign of the number with the higher absolute value.

    Examples: $(-1) + (3) = 2$
                   $7 + (-9) = -2$

SUBTRACTION OF SIGNED NUMBERS

- Change the sign of the number that is being subtracted. Then add the numbers.

  Examples: $(-2) - (5) = (-2) + (-5) = -7$

  $(-3) - (-4) = (-3) + (+4) = 1$

  $(7) - (8) = (7) + (-8) = -1$

−1, 0, 1 AND NUMBERS IN BETWEEN

- $a + 0 = a$ $a \cdot 0 = 0$ $\dfrac{a}{0}$ is undefined $\dfrac{0}{a} = 0$ as long as $a \neq 0$
- $a \cdot 1 = a$ $\dfrac{a}{1} = a$
- $a \cdot -1 = -a$ $\dfrac{a}{-1} = -a$

- The **reciprocal** of a number is 1 divided by that number. The product of a number and its reciprocal equals 1. The reciprocal of 3 is $\dfrac{1}{3}$. The reciprocal of $-\dfrac{2}{3}$ is $-\dfrac{3}{2}$. The reciprocal is also called the multiplicative inverse.

- When you multiply any positive number by a number between 0 and 1, the number gets smaller.

  Examples: $\dfrac{1}{2} \cdot 6 = 3$

  $\dfrac{1}{2} \cdot \dfrac{3}{5} = \dfrac{3}{10}$ → $\dfrac{3}{10}$ is less than both $\dfrac{1}{2}$ and $\dfrac{3}{5}$.

- The reciprocals of numbers between 0 and 1 are greater than the original numbers.

  Example: The reciprocal of $\dfrac{4}{5}$ is $\dfrac{5}{4}$. $\dfrac{5}{4} > \dfrac{4}{5}$

- The squares of numbers between 0 and 1 are less than the original numbers.

  Example: The square of $\dfrac{2}{5}$ is $\dfrac{4}{25}$. $\dfrac{4}{25} < \dfrac{2}{5}$

- If $0 > \dfrac{a}{b} > 1$, then $\dfrac{a}{b} > \left(\dfrac{a}{b}\right)^2 > \left(\dfrac{a}{b}\right)^3$

  Example: $\dfrac{1}{2} > \left(\dfrac{1}{2}\right)^2 > \left(\dfrac{1}{2}\right)^3$ → $\dfrac{1}{2} > \dfrac{1}{4} > \dfrac{1}{8}$

> **REMEMBER**
>
> All positive powers of numbers between 0 and 1 are less than the original numbers.

- The reciprocal of numbers between −1 and 0 are less than the original numbers.

  Example: The reciprocal of $-\dfrac{4}{5}$ is $-\dfrac{5}{4}$. $-\dfrac{5}{4} < -\dfrac{4}{5}$

- The squares of numbers between −1 and 0 are greater than the original numbers.

  Example: The square of $-\dfrac{2}{5}$ is $\dfrac{4}{25}$. $\dfrac{4}{25} > -\dfrac{2}{5}$

- If $-1 > \dfrac{c}{d} > 0$, then $\left(\dfrac{c}{d}\right)^2 > \left(\dfrac{c}{d}\right)^3 > \dfrac{c}{d}$

  Example: $\left(-\dfrac{1}{2}\right)^2 > \left(-\dfrac{1}{2}\right)^3 > -\dfrac{1}{2}$ → $\dfrac{1}{4} > -\dfrac{1}{8} > -\dfrac{1}{2}$

ORDER OF OPERATIONS (PEMDAS)

PEMDAS stands for

- Parentheses
- Exponents
- Multiplication and Division
- Addition and Subtraction

INTRODUCTION TO EXPONENTS

- When a number k is multiplied by itself n times, it is represented as k^n. k is called the **base**, and n is called the **exponent**.

$$2 \cdot 2 \cdot 2 \cdot 2 \cdot 2 = 2^5 \qquad k \cdot k \cdot k \cdot k = k^4$$

- $x^1 = x \qquad 42^1 = 42 \qquad 7^1 = 7$
- $x^0 = 1 \qquad 13^0 = 63^0 = 2^0 = 1$

Example: $((5^2 - 22)^3 \div 9 - 2)^{66}$ To start, calculate 5^2.

$((25 - 22)^3 \div 9 - 2)^{66}$ Subtract inside the parentheses.

$((3)^3 \div 9 - 2)^{66}$ Determine the inner exponent, $3^3 = 27$.

$(27 \div 9 - 2)^{66}$ Divide first.

$(3 - 2)^{66} = 1^{66} = 1$ Work inside the parentheses and then calculate the exponent.

- If there are consecutive multiplications and divisions in an expression, perform them from left to right.

Example: $50 \div 2 \times 5 = 25 \times 5 = 125$ not $50 \div 10 = 5$

See Section 3.3 for complete coverage of exponents.

SAMPLE PROBLEMS

EXAMPLE 1

$$\frac{-3(-5)^2 - 7 \cdot 2^3 + 7 \cdot 3}{\frac{3}{2} \cdot 14 - |2 - 12|} = ?$$

$\dfrac{-3(-5)^2 - 7 \cdot 2^3 + 7 \cdot 3}{\frac{3}{2} \cdot 14 - |2 - 12|}$ Work the exponents first.

You can also perform the subtraction inside the absolute value since absolute value signs are treated as parentheses.

$\dfrac{-3 \cdot 25 - 7 \cdot 8 + 7 \cdot 3}{\frac{3}{2} \cdot 14 - |-10|}$ Perform all multiplications, and replace $|-10|$ with 10.

$\dfrac{-75 - 56 + 21}{21 - 10}$ Perform all additions and subtractions.

$\dfrac{-75 - 56 + 21}{21 - 10} = \dfrac{-110}{11} = -10$

EXAMPLE 2

The product of four different prime numbers is an even number. Which of the following is the least of the four numbers?

    (A) 1   (B) 2   (C) 3   (D) 5   (E) 7

All prime numbers except 2 are odd. An odd number times an odd number is always odd. So if the overall product is even, there has to be at least one even number in the mix. Therefore, the least number of the four is 2. Remember that 1 is not a prime number.

The answer is (B).

EXAMPLE 3

If $0 < b < a$ which of the following is always true?

(A) $\frac{b}{a} > 1$　　(B) $\frac{b^2}{a^2} > \frac{b}{a}$　　(C) $\frac{b}{a} > \frac{b^2}{a^2}$　　(D) $\frac{b^3}{a^3} > \frac{b}{a}$　　(E) $-b < -a$

If $0 < b < a$, then $\frac{b}{a} < 1$. For example, if $a = 3$ and $b = 2$, then $\frac{2}{3} < 1$.

(A) $\frac{b}{a} > 1$ is false because when $\frac{a}{b} > 1$, $\frac{b}{a} < 1$.

(B) $\frac{b^2}{a^2} > \frac{b}{a}$ is false because when $\frac{b}{a} < 1$, its square is less than itself.

(C) $\frac{b}{a} > \frac{b^2}{a^2}$ is true. You can try numbers to verify your answer. If $a = 3$ and $b = 2$, then $\frac{2}{3} > \frac{4}{9}$.

(D) $\frac{b^3}{a^3} > \frac{b}{a}$ is false. Since $\frac{b}{a} < 1$, it gets smaller as you take higher powers.

(E) $-b < -a$ is false. Since $b < a$, $-b > -a$. For example, if $a = 3$ and $b = 2$, then $-2 > -3$.

The answer is (C).

EXAMPLE 4

If $n + 1$ is an odd number, which of the following must be an odd number?

(A) n　　(B) $2(n + 1)$　　(C) n^n　　(D) $n^2 + n$　　(E) $(n + 1)^n$

If $n + 1$ is an odd number, n must be an even number. For example, if $n + 1 = 5$ then $n = 4$. Assume $n = 4$ and work your way through the answer choices.

(A) 4 　　　　　　　　　　　　　　　　(even)
(B) $2(4 + 1) = 10$ 　　　　　　　　　(even)
(C) $4^4 = 4 \cdot 4 \cdot 4 \cdot 4$ 　　　　　　　(even)
(D) $4^2 + 4 = 20$ 　　　　　　　　　(even)
(E) $(4 + 1)^4 = 5 \cdot 5 \cdot 5 \cdot 5 = 625$ 　(odd)

The answer is (E).

Factors, Multiples, Divisibility, and Remainders

FACTORS AND MULTIPLES

- The factors of a number are positive integers that divide the number evenly.

  Example: The factors of 24 are 24, 12, 8, 6, 4, 3, 2, and 1.

- Multiples of any given number are those numbers that can be divided by that number evenly.

  Example: 30, 45, 60, and 75 are all multiples of 15.

DIVISIBILITY

- A number is divisible by another number if the result (quotient) is an integer. The remainder is 0.

  Example: $32 = 8 \cdot 4$; therefore, 4 and 8 are factors of 32. So 32 is divisible by 4, and 32 is divisible by 8.

- If a number is divisible by two different prime numbers, then it must be divisible by the product of those prime numbers.

  Example: 36 is divisible by 2, and it is divisible by 3. Therefore it is divisible by 6.

- If a number A is divisible by a number B, then A is also divisible by all factors of B.

  Examples: 28 is divisible by 14. Therefore, it is also divisible both by 7 and 2 ($14 = 2 \cdot 7$). 48 is divisible by 24. Therefore, it is also divisible by 2, 3, 4, 6, 8, and 12.

- Below is a list of common divisibility rules.

  2 → All even numbers are divisible by 2.

  3 → If the sum of the digits of a number is divisible by 3, the number is also divisible by 3.

  Example: To check if 1,458 is divisible by 3, add $1 + 4 + 5 + 8 = 18$. 18 is divisible by 3, so 1,458 is also divisible by 3.

  4 → If the last two digits of a number are 00 or are divisible by 4, then the number is divisible by 4.

  Example: 23,456 is divisible by 4 since 56 is divisible by 4.

  Alternatively, divide the last two digits by 2. If the result is even, then the number is divisible by 4. Since the number is divisible by 2 twice, it is divisible by 4.

  5 → If the units digit of a number is 0 or 5, then the number is divisible by 5.

  6 → If a number is divisible by both 2 and 3, then the number is divisible by 6.

  Example: 546 is divisible by 6 because it is even (divisible by 2) and is divisible by 3 ($5 + 4 + 6 = 15$, 15 is divisible by 3).

  9 → If the sum of the digits of a number is divisible by 9, the number is also divisible by 9.

  Example: To check if 4,608 is divisible by 9, add $4 + 6 + 0 + 8 = 18$. 18 is divisible by 9, so 4,608 is also divisible by 9.

PRIME FACTORIZATION

- Every integer greater than 1 can be expressed as a product of a set of prime numbers.

  Examples: $30 = 2 \cdot 3 \cdot 5$
  $124 = 2 \cdot 2 \cdot 31 = 2^2 \cdot 31$

- Prime factorization is finding which prime numbers multiply together to result in the original number. To find the prime factors of a number, start by dividing the original number by a prime number. Write the result as the branches of a tree as shown on the right, $108 = 3 \cdot 36$. Keep moving down until you are left with prime numbers only.

  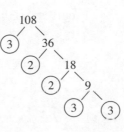

  $$108 = 2 \cdot 2 \cdot 3 \cdot 3 \cdot 3 = 2^2 \cdot 3^3$$

- To find the total number of all positive factors of an integer, first use prime factorization. If $A = a^m \cdot b^n \cdot c^t$, the total number of all positive factors of A is $(m + 1)(n + 1)(t + 1)$.

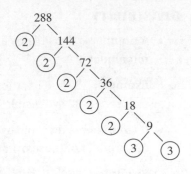

  Example: $288 = 2^5 \cdot 3^2$, so the total number of all positive factors of 288 is $(5 + 1)(2 + 1) = 18$.

LEAST COMMON MULTIPLE "LCM" (OR LEAST COMMON DENOMINATOR IN FRACTIONS)

- To find the LCM of two numbers, find the prime factorization of each number. Then multiply one of each common factor and all of the factors that are not common.
- Organizing the prime factors as shown below is a simple way to see the factors that go into the LCM.

  Examples:

24	$= 2 \cdot 2 \cdot 2 \cdot 3$
32	$= 2 \cdot 2 \cdot 2 \cdot \quad 2 \cdot 2$
LCM	$= 2 \cdot 2 \cdot 2 \cdot 3 \cdot 2 \cdot 2 = 96$

120	$= 2 \cdot 2 \cdot 2 \cdot 3 \cdot 5$
48	$= 2 \cdot 2 \cdot 2 \cdot 3 \quad\quad 2$
LCM	$= 2 \cdot 2 \cdot 2 \cdot 3 \cdot 5 \cdot 2 = 240$

REMAINDER PROBLEMS

- A remainder is the integer left over when you divide two numbers. It is not the decimal part of the quotient.

  $$\begin{array}{r} 5 \quad R = 2 \\ 3\overline{)17} \end{array} \qquad 17 \div 5 = 3, \text{ remainder} = 2 \qquad \begin{array}{r} \text{quotient} \quad R = \text{remainder} \\ \text{divisor}\,\overline{)\,\text{dividend}} \end{array}$$
  $$17 = (3 \cdot 5) + 2$$

- When an integer is divided by an integer larger than itself (for example $9 \div 12$), the quotient is zero and the remainder is the smaller integer (quotient = 0, remainder = 9).

COUNTING PROBLEMS

- Inclusive means the two endpoints of a range are included in the set.

  Example: How many integers are between 75 and 152 inclusive? $152 - 75 + 1 = 78$

  Inclusive: Final Number − Initial Number + 1

- Exclusive means the two endpoints of a range are not included in the set.

  Example: How many integers are between 46 and 112 exclusive? $112 - 46 - 1 = 65$

  Exclusive: Final Number − Initial Number − 1

PLACE VALUE

- The value of a digit depends on its place in a number. The places are named as follows:

thousands digit (8 · 1,000)	hundreds digit (2 · 100)	tens digit (3 · 10)	ones (units) digit (4 · 1)	decimal point	tenths digit (4 · 0.1)	hundredths digit (0 · 0.01)	thousandths digit (3 · 0.001)
8	2	3	4	.	4	0	3

$$8{,}234.403 = (8 \cdot 1{,}000) + (2 \cdot 100) + (3 \cdot 10) + (4 \cdot 1) + (4 \cdot 0.1) + (0 \cdot 0.01) + (3 \cdot 0.001)$$

Example: $43.147 = (4 \cdot 10) + (3 \cdot 1) + (1 \cdot 0.1) + (4 \cdot 0.01) + (7 \cdot 0.001)$

- The place value of a number is the product of the number by the value of its position.

The place value of 3 in 365 is $3 \cdot 100 = 300$
The place value of 5 in 1,365 is $5 \cdot 1 = 5$
The place value of 6 in 0.365 is $6 \cdot 0.01 = 0.06$

Example: If P and Q represent digits, the value of PQ can be calculated as $PQ = 10P + Q$. Similarly,

$$QP = 10Q + P$$
$$PPQ - 100P + 10P + Q$$
$$335 = (3 \cdot 100) + (3 \cdot 10) + (5 \cdot 1)$$

ROUNDING

- To round a number to a specific digit:

1. Check the digit to the right of that specific digit.
2. If the digit to the right is 5 or more, round your digit up one. If it is less than 5, make no change.
3. If you are in the decimal places, eliminate all digits to the right of the specific digit. If you are rounding the whole part, replace all numbers to the right with 0s.

Examples: Round 23.058 to the nearest hundredths.

The hundredths digit of the number is 5, and the digit to the right of it is 8. So round 5 up to 6, and eliminate 8.

$23.058 \approx 23.06$

Round 34,643 to the nearest hundred.

The hundredths digit is 6, and the digit to the right of it is 4. Make no change to 6, and replace all numbers to the right with zeros.

$34{,}643 \approx 34{,}600$

EXAMPLE 1

If $\frac{18}{n}$ is an integer, how many different integer values can n have?

(A) 4

(B) 5

(C) 6

(D) 10

(E) 12

Write 18 as a multiple of its prime factors using prime factorization:
$18 = 2 \cdot 3 \cdot 3 = 2 \cdot 3^2$

$1 \cdot 18 = 18$
$2 \cdot 9 \ = 18$
$3 \cdot 6 \ = 18$

The positive factors of 18 are 18, 9, 6, 3, 2, and 1. The question does not specify that the result is a positive integer, so we need to count the negative values as well, -18, -9, -6, -3, -2, and -1. Therefore n can have 12 different values.

Alternatively, use the formula given earlier. If $18 = 2^1 \cdot 3^2$, then the number of all positive factors of 18 equals $(1 + 1)(2 + 1) = 6$. Since the negatives of all those numbers will make $\frac{18}{n}$ an integer as well, there are $6 \cdot 2 = 12$ values for n.

The answer is (E).

EXAMPLE 2

How many digits are used to number the pages of a 250 page book?

(A) 250

(B) 251

(C) 640

(D) 642

(E) 750

The question is asking us to count each digit. There are nine 1-digit numbers, ninety 2-digit numbers (99 minus 9) and 151 (250 minus 99) 3-digit numbers. Therefore:

$1 \cdot 9 \ \ = \ \ \ \ 9$
$2 \cdot 90 \ = 180$
$3 \cdot 151 = 453$
Total $\ = 642$

The answer is (D).

EXAMPLE 3

> When x is divided by 5, the remainder is 2. Which of the following will result in a remainder of 4 when divided by 5?
>
> (A) $x + 1$
> (B) $x + 3$
> (C) $x + 5$
> (D) $x + 7$
> (E) $x + 9$

The easiest way to solve remainder problems is to find a number that works for the question. Since the remainder is 2 when x is divided by 5, we could use 7 or 12 for x. Let's stick with the smaller option, 7.

(A) $7 + 1 = 8$, remainder $= 3$
(B) $7 + 3 = 10$, remainder $= 0$
(C) $7 + 5 = 12$, remainder $= 2$
(D) $7 + 7 = 14$, remainder $= 4$
(E) $7 + 9 = 16$, remainder $= 1$

The answer is (D).

EXAMPLE 4

> If $60 \cdot n$ is the square of an integer, what is the least possible value that n could have?
>
> (A) 6
> (D) 9
> (C) 12
> (D) 15
> (E) 60

Write 60 as a multiple of its prime factors: $60 = 2 \cdot 2 \cdot 3 \cdot 5 = 2^2 \cdot 3 \cdot 5$.

2^2 is already a perfect square. If we multiply 60 by another 3 and by another 5, then all prime numbers will be perfect squares $\rightarrow 2 \cdot 2 \cdot 3 \cdot 5 \cdot 3 \cdot 5 = 2^2 \cdot 3^2 \cdot 5^2$.

The least number we could multiply it by is 15 ($3 \cdot 5$). The number 60 also works, but 15 is less than 60.

The answer is (D).

Fractions

- A fraction stands for the division or the ratio of two numbers.

  $a \div b$, $\dfrac{a}{b}$, and $a : b$ all mean "a divided by b."

  $$\dfrac{a \to \text{numerator}}{b \to \text{denominator}}$$

  $\dfrac{27}{1} = 27$ $\qquad$ $\dfrac{0}{1} = 0$ $\qquad$ $\dfrac{5}{0} = \text{undefined}$ $\qquad$ $\dfrac{x}{x} = \dfrac{17}{17} = \dfrac{-8}{-8} = \dfrac{142}{142} = 1$

EQUIVALENT FRACTIONS

- If both the numerator and the denominator of a fraction are multiplied (or divided) by the same nonzero number, the value of the fraction does not change. The resulting fraction and the original fraction are equivalent. In other words, they have the same value.

  $$\frac{a}{b} = \frac{a \cdot c}{b \cdot c} = \frac{a \div k}{b \div k}$$

  Example:

  $$\frac{12}{18} = \frac{12 \div 6}{18 \div 6} = \frac{2}{3} = \frac{2 \cdot 5}{3 \cdot 5} = \frac{10}{15}$$

  $\dfrac{12}{18}$, $\dfrac{2}{3}$, and $\dfrac{10}{15}$ are equivalent fractions.

MIXED NUMBERS

- Mixed numbers contain an integer and a fraction portion. $3\frac{2}{3}$ is 3 and $\frac{2}{3}$, which is essentially $3 + \frac{2}{3}$.
- To convert a mixed number into a fraction, multiply the denominator by the whole number and add that product to the numerator.

  $$5\frac{2}{3} = \frac{(5 \cdot 3) + 2}{3} = \frac{17}{3}$$

  then add · first multiply

- To convert a fraction into a mixed number, divide the numerator by the denominator. The quotient becomes the whole number portion, and the remainder becomes the new numerator. Note that the denominator never changes.

  $\dfrac{19}{5}$ $\qquad$ $5\overline{)19}\;\;\dfrac{3 \quad R = 4}{}$ $\qquad$ $\dfrac{19}{5} = 3\dfrac{4}{5}$

REDUCING FRACTIONS

- Reducing a fraction means dividing the numerator and the denominator by the same nonzero number until they have no common factors. The resulting fraction is said to be reduced to lowest terms.

  Example:

  $$\frac{72}{48} = \frac{72 \div 2}{48 \div 2} = \frac{36}{24} = \frac{36 \div 6}{24 \div 6} = \frac{6}{4} = \frac{6 \div 2}{4 \div 2} = \frac{3}{2}$$

- You can divide by any nonzero number in any order as long as you divide the numerator and the denominator by the same nonzero number.
- GMAT questions will require you to reduce fractions as much as possible to save time and effort. In addition, the answer choices will always be given in their lowest terms.

ADDITION AND SUBTRACTION OF FRACTIONS

- To add or subtract two fractions with the same denominator, add or subtract the numerators and keep the denominator.

$$\frac{a}{b} + \frac{c}{b} = \frac{a+c}{b}$$

$$\frac{a}{b} - \frac{c}{b} = \frac{a-c}{b}$$

Examples: $\frac{2}{5} + \frac{7}{5} = \frac{9}{5} = 1\frac{4}{5}$

$$\frac{7}{15} - \frac{9}{15} = -\frac{2}{15}$$

- To add/subtract two fractions with different denominators:

  1. Start by finding the LCM of the two denominators.
  2. Expand each fraction so they all have a common denominator.
  3. Add or subtract the numerators, and keep the denominator.

$\frac{2}{14} + \frac{5}{21}$ Find the least common multiple of 14 and 21.

$$\begin{array}{rl} 14 & = 2 \cdot 7 \\ 21 & = \; 7 \cdot 3 \\ \hline \text{LCM} & = 2 \cdot 7 \cdot 3 = 42 \end{array}$$

Expand both fractions to make their denominators 42.

$\frac{2}{14} \cdot \frac{3}{3} + \frac{5}{21} \cdot \frac{2}{2} = \frac{6}{42} + \frac{10}{42} = \frac{16}{42}$ Simplify.

$$\frac{16 \div 2}{42 \div 2} = \frac{8}{21}$$

MULTIPLICATION OF FRACTIONS

- To multiply two fractions, multiply numerators by numerators and multiply denominators by denominators.

$$\frac{a}{c} \cdot \frac{b}{d} = \frac{a \cdot b}{c \cdot d}$$

Example: $\frac{2}{7} \cdot \frac{14}{3} \cdot \frac{6}{5} = \frac{168}{105}$

Divide both the numerator and denominator by 21.

$$\frac{168 \div 21}{105 \div 21} = \frac{8}{5} = 1\frac{3}{5}$$

Note: A better approach is to simplify before you multiply. When multiplying fractions, any numerator can be simplified by any denominator.

Example: $\frac{2}{\cancel{7}_1} \cdot \frac{\cancel{14}^2}{\cancel{3}_1} \cdot \frac{\cancel{6}^2}{5} = \frac{2}{1} \cdot \frac{2}{1} \cdot \frac{2}{5} = \frac{8}{5} = 1\frac{3}{5}$

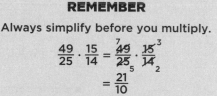

REMEMBER
Always simplify before you multiply.
$$\frac{49}{25} \cdot \frac{15}{14} = \frac{\cancel{49}^7}{\cancel{25}_5} \cdot \frac{\cancel{15}^3}{\cancel{14}_2}$$
$$= \frac{21}{10}$$
$$= 2.1$$

DIVISION OF FRACTIONS

- To divide two fractions, take the reciprocal of (flip) the divisor (the second fraction) and multiply.

$$\frac{a}{b} \div \frac{c}{d} = \frac{a}{b} \cdot \frac{d}{c} = \frac{a \cdot d}{b \cdot c}$$

Examples:

$$\frac{11}{3} \div \frac{5}{6} = \frac{11}{3} \cdot \frac{6^2}{5} = \frac{11}{1} \cdot \frac{2}{5} = \frac{22}{5} = 4\frac{2}{5}$$

$$\frac{3}{7} \div 12 = \frac{3^1}{7} \cdot \frac{1}{12_4} = \frac{1}{7} \cdot \frac{1}{4} = \frac{1}{28}$$

$$13 \div \frac{1}{3} = \frac{13}{1} \cdot \frac{3}{1} = 39$$

$$\frac{\frac{4}{5}}{\frac{12}{25}} = \frac{4}{5} \div \frac{12}{25} = \frac{4^1}{5_1} \cdot \frac{25^5}{12_3} = \frac{1}{1} \cdot \frac{5}{3} = 1\frac{2}{3}$$

Note: To multiply or divide mixed numbers, convert them into fractions first.

$$5\frac{1}{9} \div \frac{1}{6} = \frac{46}{9} \div \frac{1}{6} = \frac{46}{9} \cdot \frac{6^2}{1} = \frac{46}{3} \cdot \frac{2}{1} = \frac{92}{3} = 30\frac{2}{3}$$

CONVERTING A FRACTION INTO A DECIMAL

- To convert a fraction into a decimal, simply divide the numerator by the denominator.

Example: $\frac{4}{5} = 0.8$

$$\begin{array}{r} 0.8 \\ 5\overline{)4.0} \\ \underline{4.0} \\ 0 \end{array}$$

CONVERTING A DECIMAL INTO A FRACTION

- To convert a decimal into a fraction, remove the decimal point and divide by a power of 10 based on the number of decimal places. For 1 decimal place, divide by 10. For 2 decimal places, divide by 100. For 3 decimal places, divide by 1,000 and so on. Then reduce the fraction.

0.42   2 decimal places, so divide 42 by 10^2 (100) $\frac{42}{100} = \frac{21}{50}$

1.025   3 decimal places, so divide 1,025 by 10^3 (1,000) $\frac{1,025}{1,000} = \frac{41}{40}$

"SPLITTING" FRACTIONS

- If a fraction has one term in its denominator and an addition or subtraction in its numerator, it can be written as the addition or subtraction of two fractions.

$$\frac{a + b}{c} = \frac{a}{c} + \frac{b}{c}$$

Example: $\frac{23 - x}{x} = \frac{23}{x} - \frac{x}{x} = \frac{23}{x} - 1$

- If the denominator is a sum or difference of two terms, you cannot split the fraction.

$$\frac{a}{b + c} \neq \frac{a}{b} + \frac{a}{c}$$

Example: $\frac{12}{12 - x} \neq \frac{12}{12} - \frac{12}{x}$

COMPLEX FRACTIONS

- A complex fraction has a fraction in the numerator and another in the denominator.
- When simplifying complex fractions, start by carrying out the operations in the numerator separately from those in the denominator.

SAMPLE PROBLEMS

EXAMPLE 1

$$\dfrac{\frac{2}{3}+\frac{1}{6}}{\frac{4}{7}\cdot\frac{3}{2}}= ?$$

$\dfrac{\frac{2}{3}+\frac{1}{6}}{\frac{4}{7}\cdot\frac{3}{2}}$ First perform the operations in the numerator and in the denominator of the complex fraction.

$\dfrac{\frac{4}{6}+\frac{1}{6}}{\frac{12}{14}}$ Make the denominators equal in the numerator by multiplying the first fraction by $\frac{2}{2}$.

Multiply the fractions on the bottom.

$\dfrac{\frac{5}{6}}{\frac{12}{14}}$ To divide the fractions, flip the bottom fraction and multiply.

$\dfrac{5}{\cancel{6}_{3}}\cdot\dfrac{\cancel{14}^{7}}{12}=\dfrac{35}{36}$ Simplify before you multiply.

EXAMPLE 2

$$\dfrac{\frac{1}{9}+\frac{1}{6}}{\frac{4}{9}-\frac{3}{2}}= ?$$

Solution 1:

$\dfrac{\frac{1}{9}+\frac{1}{6}}{\frac{4}{9}-\frac{3}{2}}$ First carry out the operations in the numerator and in the denominator of the complex fraction.

$\dfrac{\frac{1}{9}\cdot\frac{2}{2}+\frac{1}{6}\cdot\frac{3}{3}}{\frac{4}{9}\cdot\frac{2}{2}-\frac{3}{2}\cdot\frac{9}{9}}$ Expand the fractions to make the denominators equal.

$\dfrac{\frac{2}{18}+\frac{3}{18}}{\frac{8}{18}-\frac{27}{18}}$ Add the fractions in the numerator, and subtract those in the denominator.

$\dfrac{\frac{5}{18}}{\frac{-19}{18}}$ To divide the fractions, flip the bottom fraction and multiply.

$\dfrac{5}{\cancel{18}_{1}}\cdot\dfrac{\cancel{18}^{1}}{-19}=-\dfrac{5}{19}$ Simplify.

Solution 2:

$$\frac{\frac{1}{9} + \frac{1}{6}}{\frac{4}{9} - \frac{3}{2}}$$

First find the least common multiple of all denominators:
LCM (2, 3, 6, 9) = 18.

$$\frac{\left(\frac{1}{9} + \frac{1}{6}\right) \cdot 18}{\left(\frac{4}{9} - \frac{3}{2}\right) \cdot 18}$$

Multiply both the numerator and the denominator by 18, and distribute the 18.

$$\frac{\frac{18}{9} + \frac{18}{6}}{\frac{4 \cdot 18}{9} - \frac{3 \cdot 18}{2}} = \frac{2 + 3}{8 - 27} = -\frac{5}{19}$$

COMPARING TWO FRACTIONS

- If two positive fractions have the same denominator, the one with the greater numerator is larger.

$$\frac{8}{17} > \frac{5}{17}$$

- In general, to compare fractions, make all denominators the same. The fraction with the greatest numerator is the largest.

$$a = \frac{2}{3} \qquad b = \frac{3}{5} \qquad c = \frac{4}{7}$$

$$a = \frac{2}{3} = \frac{70}{105} \qquad b = \frac{3}{5} = \frac{63}{105} \qquad c = \frac{4}{7} = \frac{60}{105}$$

$$\frac{70}{105}, \frac{63}{105}, \frac{60}{105} \qquad \rightarrow \qquad a > b > c$$

- If two positive fractions have the same numerator, the one with the smaller denominator is greater.

$$\frac{8}{17} > \frac{8}{21}$$

- If a positive fraction has a greater numerator and a smaller denominator than another positive fraction, the former is greater.

$$\frac{8}{17} > \frac{5}{19}$$

- If only two fractions are to be compared, you can cross multiply.

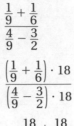

When comparing $\frac{9}{14}$ to $\frac{2}{3}$, multiply 3 and 9, and write 27 on the left. Multiply 14 and 2, and write 28 on the right.

$$\frac{9}{14} \quad < \quad \frac{2}{3}$$

Since 28 is greater than 27, $\frac{9}{14} < \frac{2}{3}$.

SAMPLE PROBLEMS

EXAMPLE 1

$$\frac{\frac{3}{2} - \frac{4}{7}}{1\frac{5}{7} + \frac{9}{14}} \cdot \frac{11}{26} = ?$$

$$\frac{\frac{3}{2} - \frac{4}{7}}{1\frac{5}{7} + \frac{9}{14}} \cdot \frac{11}{26}$$

Convert the mixed number into a fraction.
Find the LCM of the denominators: LCM(2, 7, 14) = 14.

$$\frac{\left(\frac{3}{2} - \frac{4}{7}\right) \cdot 14}{\left(\frac{12}{7} + \frac{9}{14}\right) \cdot 14} \cdot \frac{11}{26}$$

Multiply both the numerator and the denominator of the complex fraction by 14.
Distribute the 14 and simplify.

$$\frac{\frac{3}{2} \cdot 14 - \frac{4}{7} \cdot 14}{\frac{12}{7} \cdot 14 + \frac{9}{14} \cdot 14} \cdot \frac{11}{26}$$

Simplify the denominator and the numerator of the complex fraction.

$$\frac{\frac{3 \cdot 14}{2} - \frac{4 \cdot 14}{7}}{\frac{12 \cdot 14}{7} + \frac{9 \cdot 14}{14}} \cdot \frac{11}{26} = \frac{21 - 8}{24 + 9} \cdot \frac{11}{26} = \frac{13}{33} \cdot \frac{11}{26} = \frac{1 \cdot 1}{3 \cdot 2} = \frac{1}{6}$$

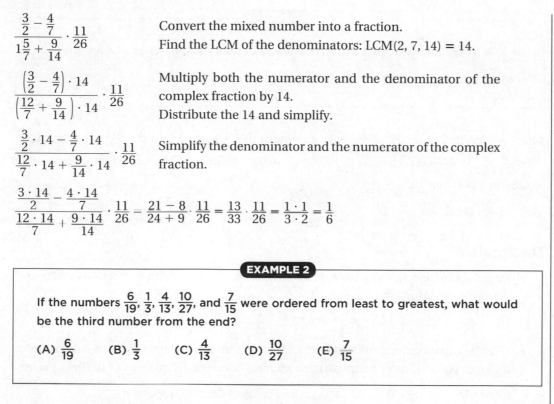

EXAMPLE 2

If the numbers $\frac{6}{19}, \frac{1}{3}, \frac{4}{13}, \frac{10}{27}$, and $\frac{7}{15}$ were ordered from least to greatest, what would be the third number from the end?

(A) $\frac{6}{19}$ (B) $\frac{1}{3}$ (C) $\frac{4}{13}$ (D) $\frac{10}{27}$ (E) $\frac{7}{15}$

The third number from the end is the middle number.
If you look closely, all the fractions are very close to $\frac{1}{3}$.
An efficient way to compare all five fractions is to compare each fraction to $\frac{1}{3}$.

- To compare $\frac{6}{19}$, expand $\frac{1}{3} = \frac{6}{18}$. Since they have the same numerator, $\frac{6}{19} < \frac{6}{18} = \frac{1}{3}$.
  $\frac{6}{19}$ is less than $\frac{1}{3}$.

- To compare $\frac{4}{13}$, expand $\frac{1}{3} = \frac{4}{12}$. Since they have the same numerator, $\frac{4}{13} < \frac{4}{12} = \frac{1}{3}$.
  $\frac{4}{13}$ is less than $\frac{1}{3}$.

- To compare $\frac{10}{27}$, expand $\frac{1}{3} = \frac{9}{27}$. Since they have the same denominator, $\frac{10}{27} > \frac{9}{27} = \frac{1}{3}$.
  $\frac{10}{27}$ is greater than $\frac{1}{3}$.

- To compare $\frac{7}{15}$, expand $\frac{1}{3} = \frac{5}{15}$. Since they have the same denominator, $\frac{7}{15} > \frac{5}{15} = \frac{1}{3}$.
  $\frac{7}{15}$ is greater than $\frac{1}{3}$.

Two of the fractions are to the right of (greater than) $\frac{1}{3}$, and two of them are to the left of (less than) $\frac{1}{3}$.
Therefore, the middle fraction is $\frac{1}{3}$.
The answer is (B).

Note: In this case, you did not need to compare the two smaller fractions, $\frac{4}{13}$ and $\frac{6}{19}$.

If need be, you can compare fractions by cross multiplication $\overset{76}{\underset{}{\frac{4}{13}}} \diagdown \overset{78}{\underset{}{\frac{6}{19}}}$.

Therefore, $\frac{4}{13} < \frac{6}{19}$.

If $\frac{n}{21}$ is a number between $\frac{2}{3}$ and $\frac{2}{7}$, how many of the possible values of n are prime?

(A) 3   (B) 5   (C) 7   (D) 9   (E) 11

Make all denominators equal for easy comparison: $\frac{2}{3} \cdot \frac{7}{7} = \frac{14}{21}$ and $\frac{2}{7} \cdot \frac{3}{3} = \frac{6}{21}$. Therefore, $\frac{6}{21} < \frac{n}{21} < \frac{14}{21}$. So n could be 7, 8, 9, 10, 11, 12, or 13. The prime numbers among the possible values are 7, 11, and 13.

The answer is (A).

Decimals

REMEMBER

Adding zeros to the right of the decimal does not affect its value.
0.03 = 0.030
     = 0.0300

- Decimals are another way to represent fractions. Think of them as fractions with denominators that are multiples of 10.

  $$0.7 = \frac{7}{10} \qquad 0.12 = \frac{12}{10^2} = \frac{12}{100} \qquad 0.009 = \frac{9}{10^3} = \frac{9}{1,000}$$

- When comparing decimals that have a different number of decimal points, add zeros to the right of the decimal points so each number has an equal quantity of decimal places.

  Example: Arrange 0.03, 0.023, 0.12, and 0.008 in descending order.

  Write them in a column, making sure each has an equal number of decimal places.

  0.030

  0.023

  0.120

  0.008

  This way, it is easier to see that $0.120 > 0.03 > 0.023 > 0.008$.

ADDING AND SUBTRACTING DECIMALS

- To add or subtract decimals, align along the decimal points.

  Example: $0.007 + 0.1 + 17.013 + 2.001 =$

  $$\begin{array}{r} 0.007 \\ 0.1 \\ 17.013 \\ +\ \ 2.001 \\ \hline 19.121 \end{array}$$

NOTE

To multiply a decimal by 10 quickly, move the decimal 1 place to the right.
$12.34 \cdot 10 = 123.4$

If multiplying by 100, move the decimal 2 places.
If multiplying by 1,000, move it 3 places and so on.

MULTIPLYING DECIMALS

- First multiply the decimals as if there are no decimal places. The number of decimal places in the product is equal to the total number of decimal places in the original numbers.

  Example: $0.03 \cdot 12.5 = ?$

  2 decimal places + 1 decimal place = 3 decimal places

  $125 \cdot 3 = 375$ So $0.03 \cdot 12.5 = 0.375$

Example: $0.04^2 = ?$

$0.04^2 = 0.04 \cdot 0.04$

2 decimal places + 2 decimal places = 4 decimal places

$4 \cdot 4 = 16$ So $0.04^2 = 0.0016$

DIVIDING DECIMALS

- First multiply each number by multiples of 10 until the divisor becomes an integer. Then complete the division regularly. If the dividend still has decimal places, use the same number of decimal places in the quotient.

Example: $0.125 \div 0.25 = (0.125 \cdot 100) \div (0.25 \cdot 100) = 12.5 \div 25$

$$\begin{array}{r} 0.5 \\ 25\overline{)12.5} \\ \underline{12.5} \\ 0.0 \end{array}$$

Alternatively, write the division as a fraction, expand to convert the decimals into integers, and simplify.

$$\frac{0.125}{0.25} \cdot \frac{1000}{1000} = \frac{125}{250} = \frac{1}{2} = 0.5$$

> **NOTE**
>
> To divide a decimal by 10 quickly, move the decimal 1 place to the left.
> $1.23 \div 10 = 0.123$
>
> If dividing by 100, move the decimal 2 places.
> If dividing by 1000, move it 3 places and so on.

SAMPLE PROBLEMS

EXAMPLE 1

$$\frac{0.0045 \cdot 0.09}{0.015 \cdot 0.27} = ?$$

Solution 1: Write each decimal as a fraction. Then "flip" the fractions in the denominator and simplify.

$$\frac{\frac{45}{10,000} \cdot \frac{9}{100}}{\frac{15}{1,000} \cdot \frac{27}{100}} = \frac{45}{10,000} \cdot \frac{9}{100} \cdot \frac{1,000}{15} \cdot \frac{100}{27} = \frac{1}{10} = 0.1$$

Solution 2: Multiply both the numerator and the denominator by a power of 10 to change all decimals to integers. The numerator has a total of 6 decimal places. So multiply by $10^6 = 1,000,000$.

$$\frac{0.0045 \cdot 0.09}{0.015 \cdot 0.27} \cdot \frac{1,000,000}{1,000,000} = \frac{45 \cdot 9}{15 \cdot 270} = \frac{1}{10} = 0.1$$

Solution 3: Fill in zeros to the right of each decimal as necessary to make the numerator and the denominator have the same total number of decimal places. Then eliminate the decimal places.

$$\frac{0.0045 \cdot 0.09}{0.015 \cdot 0.27} = \frac{0.0045 \cdot 0.09}{0.0150 \cdot 0.27} = \frac{45 \cdot 9}{150 \cdot 27} = \frac{1}{10} = 0.1$$

EXAMPLE 2

$$\frac{1.69}{1.3} - \frac{1.5}{20} \div \frac{0.04}{(0.4)^2} = ?$$

Remember to perform the division before subtraction (PEMDAS).

$$\frac{1.69}{1.3} - \frac{1.5}{20} \div \frac{0.04}{0.16} = \frac{1.69}{1.3} - \frac{1.5}{20} \cdot \frac{0.16}{0.04}^{4}_{1}$$

$$\frac{1.69}{1.3} - \frac{1.5}{{}_{5}20} \cdot \frac{\overset{1}{4}}{1} = \frac{1.69}{1.3} - \frac{1.5}{5}$$

Find the values of each fraction and subtract.

$$\frac{1.69}{1.3} - \frac{1.5}{5} = 1.3 - 0.3 = 1$$

SECTION 2.1—PRACTICE PROBLEMS

1. If $0 < x < 1$ which one of the following is the greatest?

   (A) x

   (B) x^2

   (C) $\sqrt{x}$

   (D) x^3

   (E) $x\sqrt{x}$

2. 12 random integers are picked among the numbers between 10 and 70. If 2 is added to the tens digit of each number, by how much does the sum of these 12 numbers increase?

   (A) 240

   (B) 120

   (C) 24

   (D) 20

   (E) 12

3. KT represents a 2-digit number. If John is KT years old and his brother is TK years old, which of the following could represent the age difference between the brothers?

   (A) 11

   (B) 16

   (C) 17

   (D) 18

   (E) It cannot be determined from the information given

4. m is a positive integer and $(m^3 + 5)^7$ is even. Which of the following must be odd?

   (A) $3m + 3$

   (B) $m^2 + 5$

   (C) $m \cdot (m + 1)$

   (D) $2m + 2m$

   (E) m^m

5. Which of the following numbers are reciprocals of each other?

    I. 5 and $\sqrt{5}$

    II. $\frac{2}{3}$ and $\frac{3}{2}$

    III. 11 and -11

    (A) I only

    (B) II only

    (C) III only

    (D) II and III only

    (E) I and II only

6. $(0.5)^2$ is how many times the number $\frac{1}{200}$?

    (A) 0.2

    (B) 8

    (C) 20

    (D) 40

    (E) 50

7. $\dfrac{0.007 \cdot 0.42}{2 \cdot 0.0049} =$

    (A) 0.3

    (B) 0.7

    (C) $\frac{1}{7}$

    (D) 3

    (E) 7

8. $\dfrac{39}{201} =$

    (A) 0.183

    (B) 0.194

    (C) 0.205

    (D) 0.251

    (E) 0.515

9. $\dfrac{1\frac{2}{7} + 2\frac{1}{3}}{3\frac{3}{5} - \frac{17}{10}} = ?$

    (A) $\frac{19}{21}$

    (B) $\frac{21}{19}$

    (C) $1\frac{19}{21}$

    (D) $\frac{76}{21}$

    (E) $\frac{76}{7}$

10. $\dfrac{0.03}{0.06 + \frac{1}{20}} \div \dfrac{3^2}{\sqrt{121}} =$

(A) 3

(B) $\dfrac{3}{27}$

(C) $\dfrac{3}{11}$

(D) $\dfrac{9}{11}$

(E) $\dfrac{1}{3}$

11. When m is divided by k, the quotient is 5 and the remainder is 26. What is the least possible value of m?

(A) 57

(B) 130

(C) 143

(D) 161

(E) 184

12. E and m are integers. When E is divided by 22, the remainder is $m - 4$ and the quotient is m. What is the least possible value of E?

(A) 18

(B) 22

(C) 26

(D) 88

(E) 111

13. If both $\dfrac{x}{6}$ and $\dfrac{48}{x}$ are positive integers, x can be how many different values?

(A) 2

(B) 3

(C) 4

(D) 5

(E) 6

14. If $\dfrac{20-x}{x}$ is a positive integer, what is the sum of all possible values of x?

(A) 5

(B) 6

(C) 21

(D) 22

(E) 42

15. When K is divided by 5, the remainder is 2. Which of the following must be divisible by 5?

(A) $2K + 2$

(B) $3K + 1$

(C) $2K - 2$

(D) $3K - 1$

(E) $5K + 1$

16. $6PQ$ is a 3-digit number (P and Q are digits) that is divisible by 4. When $6PQ$ is divided by 5, the remainder is 2. How many different values can P take?

    (A) 6
    (B) 5
    (C) 4
    (D) 3
    (E) 2

17. Amir waters his lawn once every 12 days, and Masa waters his lawn once every 15 days. In how many days will they water their lawns on the same day after the first time they water together?

    (A) 3
    (B) 15
    (C) 30
    (D) 60
    (E) 90

18. $\left(\dfrac{1}{5} - \dfrac{1}{5} \div \dfrac{4}{5} + \dfrac{1}{4}\right)^{-1} =$

    (A) $\dfrac{1}{5}$
    (B) $\dfrac{1}{4}$
    (C) 1
    (D) 4
    (E) 5

19. $\dfrac{1}{2} + \dfrac{1+\dfrac{1}{3}}{1-\dfrac{1}{3}} + \left(\dfrac{1}{6}\right)^{1} =$

    (A) $\dfrac{1}{3}$
    (B) $\dfrac{1}{2}$
    (C) $6\dfrac{1}{3}$
    (D) $8\dfrac{1}{2}$
    (E) 9

20. $\dfrac{0.02}{0.2} + \dfrac{0.0021}{0.21} + \dfrac{0.42}{420} =$

    (A) $\dfrac{1}{1110}$
    (B) $\dfrac{111}{1000}$
    (C) $\dfrac{1110}{1000}$
    (D) $\dfrac{111}{100}$
    (E) 1100

21. $x = \dfrac{3}{10}$ $y = \dfrac{5}{12}$ $z = \dfrac{12}{19}$

Which of the following must be true?

(A) $z < x < y$

(B) $y < x < z$

(C) $z < y < x$

(D) $x < z < y$

(E) $x < y < z$

22. $\left(1 + \dfrac{1}{2}\right) \cdot \left(1 + \dfrac{1}{3}\right) \cdot \left(1 + \dfrac{1}{4}\right) \cdot \ldots \cdot \left(1 + \dfrac{1}{22}\right) =$

(A) $23\dfrac{1}{2}$

(B) $23\dfrac{1}{23}$

(C) $\dfrac{23}{2}$

(D) $\dfrac{22}{23}$

(E) $\dfrac{1}{23}$

23. $\left(\dfrac{5}{3} + \dfrac{55}{33} + \dfrac{5555}{3333}\right) \cdot \dfrac{8}{5}$

(A) $\dfrac{5}{3}$

(B) $\dfrac{10}{3}$

(C) 5

(D) 8

(E) 13

24. $m = \dfrac{40}{39}$ $k = \dfrac{400}{399}$ $n = \dfrac{4000}{3999}$

Which of the following must be true?

(A) $m > k > n$

(B) $n > k > m$

(C) $m > n > k$

(D) $n > m > k$

(E) $k > m > n$

25. $\dfrac{15}{16} + \dfrac{16}{17} + \dfrac{17}{18} + \ldots + \dfrac{61}{62} + \dfrac{62}{63}$

In the addition above, each numerator and each denominator is increased by one. The new sum is how much greater than the sum above?

(A) $\dfrac{63}{64}$

(B) $\dfrac{15}{64}$

(C) $\dfrac{1}{4}$

(D) $\dfrac{3}{64}$

(E) $\dfrac{1}{64}$

26. A bank prints a 3-digit serving number ranging from 001 to 999 for each customer. Rajiv's number was 962. Bianca arrived after Rajiv and received number 016. How many customers received numbers after Rajiv and before Bianca?

    (A) 50
    (B) 51
    (C) 52
    (D) 53
    (E) 54

27. The sum of 5 different positive 2-digit integers is 130. What is the highest possible value of the largest of these integers?

    (A) 88
    (B) 84
    (C) 78
    (D) 74
    (E) 68

28. x and y are positive integers, and $x^3 = 24y$. What is the least possible value of $x + y$?

    (A) 600
    (B) 225
    (C) 18
    (D) 15
    (E) 12

29. W and V are positive integers. When W is divided by $(V - 3)$, the quotient is $(V + 3)$ and the remainder is 5. Which of the following could be the value of W?

    (A) 30
    (B) 40
    (C) 45
    (D) 55
    (E) 61

30. Letters A, B, C, D, E, F, G, and H are equally spaced on a number line in that order. If the distance between A and H is 1, the distance AG is how much longer than the distance DH?

    (A) $\dfrac{5}{7}$
    (B) 0.7
    (C) 0.5
    (D) $\dfrac{3}{7}$
    (E) $\dfrac{2}{7}$

1	2	3	4	5	6	7	8	9	10	11	12	13	14	15	16	17	18	19	20
C	A	D	E	B	E	A	B	C	E	D	D	C	D	D	B	D	E	D	B

21	22	23	24	25	26	27	28	29	30
E	C	D	A	D	C	B	D	C	E

1. **C** Pick a number that fits the definition and compare. For example, let $x = \frac{1}{4}$ (or use 0.25 if you feel more comfortable with decimals).

   (A) $x = \frac{1}{4}$

   (B) $x^2 = \frac{1}{16}$

   (C) $\sqrt{x} = \sqrt{\frac{1}{4}} = \frac{1}{2}$

   (D) $x^3 = \left(\frac{1}{4}\right)^3 = \frac{1}{64}$

   (E) $x\sqrt{x} = \frac{1}{4} \cdot \sqrt{\frac{1}{4}} = \frac{1}{4} \cdot \frac{1}{2} = \frac{1}{8}$

   Choice (C) $\frac{1}{2}$ is the greatest.

   Alternatively, compare the powers. Since $0 < x < 1$, the number with the smallest power has the greatest value.

2. **A** If 1 is added to the tens digit of an integer, its value increases by 10:

   $28 \rightarrow 38 \qquad 38 - 28 = 10.$

   If you add 2 to the tens digit, the value of the integer increases by 20:

   $28 \rightarrow 48 \qquad 48 - 28 = 20.$

   For 12 numbers, the sum increases by $12 \cdot 20 = 240$.

3. **D** If KT represents a 2-digit number, the value of this number is $10K + T$. Similarly, the value of TK is $10T + K$.

   For example, $25 = 2 \cdot 10 + 5$ and $52 = 5 \cdot 10 + 2$.

   The difference is $(10T + K) - (10K + T) = 9T - 9K = 9(T - K)$

   Therefore, the difference must be a multiple of 9. 18 is the only multiple of 9 in the answer choices.

   Alternatively, try numbers to discover the relationship.

   $21 - 12 = 9$
   $42 - 24 = 18 \checkmark$
   $51 - 15 = 36$

   9, 18, and 36 are all multiples of 9.

4. **E** If $(m^3 + 5)^7$ is even, $m^3 + 5$ must be even since all powers of odd numbers are odd.

If $m^3 + 5$ is even, then m^3 must be odd since odd + odd is an even number.

Finally, if m^3 is odd, m must be odd. Check the answer choices individually to find an odd result (see below).

Alternatively, plug in numbers and try.

Assume m is even. Plug in 2 $(m^3 + 5)^7$. $(2^3 + 5)^7 = (8 + 5)^7 = 13^7$ is odd, so m is not even.
Assume m is odd. Plug in 3 $(m^3 + 5)^7$. $(27 + 5)^7 = (32)^7$ is even. So m must be odd.
Let $m = 3$.

(A)	$3m + 3 = 3 \cdot 3 + 3 = 12$	even
(B)	$m^2 + 5 = 3^2 + 5 = 14$	even
(C)	$m(m + 1) = 3 \cdot 4 = 12$	even
(D)	$2m + 2^m = 6 + 2^3 = 6 + 8 = 14$	even
(E)	$3^3 = 27$	odd

5. **B** The reciprocal of a number A is 1 divided by A. In other words, the product of a number and its reciprocal equals 1.

Only option II contains reciprocals: $1 \div \frac{2}{3} = \frac{3}{2}$ and $\frac{2}{3} \cdot \frac{3}{2} = 1$.

6. **E** $(0.5)^2 = 0.5 \cdot 0.5 = 0.25$ and $\frac{1}{200} = 0.005$

$\frac{0.250}{0.005} = \frac{250}{5} = 50$ times

Alternatively,

$\frac{0.5^2}{\frac{1}{200}} = 0.25 \cdot \frac{200}{1} = 50$ times

You could also set up an algebraic equation:

$(0.5)^2 = n \cdot \frac{1}{200}$

$0.25 = n \cdot \frac{1}{200}$

$n = 200 \cdot 0.25 = 50$

7. **A** $\frac{0.007 \cdot 0.42}{2 \cdot 0.0049}$

One way to avoid dealing with decimals is to expand the fraction with a power of 10 so that all decimals turn into integers. In the numerator, there are a total of $3 + 2 = 5$ decimal places. In the denominator, there's only 4. Pick the larger one and multiply both the numerator and the denominator by $10^5 = 100,000$.

$\frac{0.007 \cdot 0.42}{2 \cdot 0.0049} \cdot \frac{100,000}{100,000}$

To multiply by multiples of 10 quickly, move the decimal places to the right as many times as the number of zeros.

$\frac{7 \cdot 42}{2 \cdot 490}$

Divide both the numerator and the denominator by 7 and by 2.

$\frac{21}{70} = \frac{3}{10} = 0.3$

Divide by 7 again.

Alternatively,

$$\frac{0.007 \cdot 0.42}{2 \cdot 0.0049}$$

Add zeros to the right of decimal places to make all decimals have an equal number of decimal places. If there are integers, introduce a decimal point and place zeros to the right.

$$\frac{0.0070 \cdot 0.42}{0.0049 \cdot 2.00}$$

Now that the numerator and the denominator have an equal number of total decimal places, you can remove the decimal places.

$$\frac{70 \cdot 42}{49 \cdot 200} = \frac{\overset{1}{\cancel{7}} \cdot \overset{1}{\cancel{10}} \cdot \overset{3}{\cancel{6}} \cdot \overset{1}{\cancel{7}}}{\underset{1}{\cancel{7}} \cdot \underset{1}{\cancel{7}} \cdot \underset{1}{\cancel{2}} \cdot \underset{1}{\cancel{10}} \cdot 10} = \frac{3}{10} = 0.3$$ Simplify to get 0.3.

8. **B** A straightforward way is to divide 39 by 201 using long division.

Alternatively, (and much more quickly) use approximation and elimination.

$\frac{39}{201}$ is approximately equal to $\frac{40}{200} = \frac{1}{5}$. Since the denominator of $\frac{39}{201}$ is slightly larger and its numerator is slightly smaller, $\frac{39}{201}$ must be slightly less than $\frac{1}{5}$ or 0.2. The closest answer is (B) 0.194.

9. **C** $\dfrac{1\frac{2}{7} + 2\frac{1}{3}}{3\frac{3}{5} - \frac{17}{10}} = \dfrac{\frac{9}{7} + \frac{7}{3}}{\frac{18}{5} - \frac{17}{10}}$ Convert all mixed numbers into fractions.

$$\frac{\frac{27}{21} + \frac{49}{21}}{\frac{36}{10} - \frac{17}{10}}$$

Expand the fractions, and make the denominators equal. Add the numerator, and subtract the denominator.

$$\frac{\frac{76}{21}}{\frac{19}{10}}$$

Flip the denominator, and multiply.

$$\frac{\overset{4}{\cancel{76}}}{21} \cdot \frac{10}{\underset{1}{\cancel{19}}} = \frac{4}{21} \cdot \frac{10}{1}$$ Simplify, and then multiply.

$$\frac{40}{21} = 1\frac{19}{21}$$

10. **E** $\dfrac{0.03}{0.06 + \frac{1}{20}} \div \dfrac{3^2}{\sqrt{121}}$ Convert $\frac{1}{20}$ into a decimal: $\frac{1}{20} = 0.05$. $\sqrt{121} = 11$ and $3^2 = 9$.

$$\frac{0.03}{0.06 + 0.05} \div \frac{9}{11} =$$

Add $0.06 + 0.05 = 0.11$. Flip the second fraction and multiply.

$$\frac{0.03}{0.11} \cdot \frac{11}{9} = \frac{\overset{1}{\cancel{3}}}{\cancel{11}} \cdot \frac{\overset{1}{\cancel{11}}}{\underset{3}{\cancel{9}}} = \frac{1}{3}$$

Expand the first fraction by 100 to get integers, simplify by 11, and multiply.

11. **D** $k\overline{)m}\ \ ^{5\ \ R = 26}$

$$5 \cdot k + 26 = m$$

If the remainder is 26, the least possible value of k is 27 since the divisor is always greater than the remainder.

If $k = 27$, then $m = 5 \cdot 27 + 26 = 161$.

12. **D** $22\overline{)E}\ \ \overset{m\ \ \ R = m - 4}{}$

$$22m + (m - 4) = E$$
$$23m - 4 = E$$

To minimize E, m must be as small as possible. Since the minimum value of a remainder is zero, $m - 4 = 0$. $m = 4$ is the minimum value of m.

So the least possible value of E is $23m - 4 = 23 \cdot 4 - 4 = 88$.

13. **C** From the second fraction, we conclude that x has to be a factor of 48. $48 = 2^4 \cdot 3$.

The number of positive factors $= (4 + 1)(1 + 1) = 10$.

Factors of 48:

1   48
2   24
3   16
4   12
6   8

If $\frac{x}{6}$ is a positive integer, x has to be a multiple of 6. Among the factors of 48, the multiples of 6 are 6, 12, 24, and 48.

14. **D** $\dfrac{20 - x}{x} = \dfrac{20}{x} - \dfrac{x}{x} = \dfrac{20}{x} - 1$

$\dfrac{20}{x}$ must be an integer. Therefore, x has to be a factor of 20.

The factors of 48 = {1, 2, 4, 5, 10, 20}.

$\dfrac{20}{x} - 1$ needs to be a positive integer. All factors work except 20 because $\dfrac{20}{20} - 1 - 0$ is not a positive integer. $1 + 2 + 4 + 5 + 10 = 22$

15. **D** Find a number K that fits the initial definition, such as, 7. When 7 is divided by 5, the remainder is 2. Then try the answer choices.

(A) $2 \cdot 7 + 2 = 16$ $16 \div 5 = 3$ remainder $= 1$
(B) $3 \cdot 7 + 1 = 22$ $22 \div 5 = 4$ remainder $= 2$
(C) $2 \cdot 7 - 2 = 12$ $12 \div 5 = 2$ remainder $= 2$
(D) $3 \cdot 7 - 1 = 20$ $20 \div 5 = 4$ remainder $= 0$
(E) $5 \cdot 7 + 1 = 36$ $36 \div 5 = 7$ remainder $= 1$

Alternatively, write K as $K = 5n + 2$ and plug it into the answer choices.

(A) $2(5n + 2) + 2 = 10n + 6$ Remainder $= 1$
(B) $3(5n + 2) + 1 = 15n + 7$ Remainder $= 2$
(C) $2(5n + 2) - 2 = 10n + 2$ Remainder $= 2$
(D) $3(5n + 2) - 1 = 15n + 5$ Remainder $= 0$
(E) $5(5n + 2) + 1 - 25n + 11$ Remainder $- 1$

16. **B** If the remainder is 2 when divided by 5, Q is either 2 or 7. If a number is divisible by 5, the ones digit must be 5 or 0. If the remainder is 2, the ones digit is either $5 + 2 = 7$ or $0 + 2 = 2$.

The 3-digit number is divisible by 4, so it is an even number. Therefore, Q must be 2.

If $6PQ$ is divisible by 4, the 2-digit number PQ must be divisible by 4. P could have 5 different values. The last two digits of $6PQ$ could be 12, 32, 52, 72, or 92.

17. **D** This question is simply asking for the least common multiple of 12 and 15.

$$
\begin{array}{rl}
12 = & 2 \cdot 2 \cdot 3 \\
15 = & 3 \cdot 5 \\
\hline
\text{LCM} = & 2 \cdot 2 \cdot 3 \cdot 5 = 60
\end{array}
$$

Alternatively, start writing the multiples of 12 and 15 until you find the first common multiple.

12—24—36—48—60
15—30—45—60

18. **E** $\left(\dfrac{1}{5} - \dfrac{1}{5} \div \dfrac{4}{5} + \dfrac{1}{4}\right)^{-1}$ To divide first, multiply $-\dfrac{1}{5}$ by the reciprocal of $\dfrac{4}{5}$.

$\left(\dfrac{1}{5} - \dfrac{1}{5} \cdot \dfrac{5}{4} + \dfrac{1}{4}\right)^{-1}$ Simplify and multiply.

$\left(\dfrac{1}{5} - \dfrac{1}{4} + \dfrac{1}{4}\right)^{-1}$ Add and subtract the fractions with the same denominator.

$\left(\dfrac{1}{5}\right)^{-1} = 5$ Take the reciprocal since the exponent is -1 and $a^{-1} = \dfrac{1}{a}$.

19. **D** $\dfrac{1}{2} + \dfrac{1 + \dfrac{1}{3}}{1 - \dfrac{1}{3}} + \left(\dfrac{1}{6}\right)^{-1} =$ Add and subtract the fractions in the numerator and denominator of the complex fraction first.

$\dfrac{1}{2} + \dfrac{\dfrac{4}{3}}{\dfrac{2}{3}} + \left(\dfrac{1}{6}\right)^{-1}$ To divide fractions, multiply the reciprocal of the denominator by the numerator. $\left(\dfrac{1}{6}\right)^{-1}$ with 6.

$\dfrac{1}{2} + \dfrac{4}{3} \cdot \dfrac{3}{2} + 6 = \dfrac{1}{2} + \dfrac{4}{2} + 6$

$\dfrac{1}{2} + 2 + 6 = \dfrac{1}{2} + 8 = 8\dfrac{1}{2}$

20. **B** Rewrite as $\dfrac{0.02}{0.20} + \dfrac{0.0021}{0.2100} + \dfrac{0.42}{420.00}$ Expand each fraction to convert decimals into integers.

$\dfrac{2}{20} + \dfrac{21}{2100} + \dfrac{42}{42,000}$ Reduce each fraction.

$\dfrac{1}{10} + \dfrac{1}{100} + \dfrac{1}{1000}$ Make every denominator equal 1000.

$\dfrac{100}{1000} + \dfrac{10}{1000} + \dfrac{1}{1000} = \dfrac{111}{1000}$

21. **E** If you compare each number to $\frac{1}{2}$, it is easier to notice that $\frac{12}{19}$ is the largest.

$\frac{3}{10}$ is less than $\frac{1}{2}$ because $\frac{1}{2} = \frac{5}{10} > \frac{3}{10}$.

$\frac{5}{12}$ is less than $\frac{1}{2}$ because $\frac{1}{2} = \frac{6}{12} > \frac{5}{10}$.

$\frac{12}{19} > \frac{1}{2}$ because $\frac{12}{19} \approx \frac{12}{18} = \frac{2}{3}$, which is greater than $\frac{1}{2}$.

To compare $\frac{3}{10}$ with $\frac{5}{12}$, make the denominators equal. The LCD is 60.

$\frac{3}{10} = \frac{18}{60}$ and $\frac{5}{12} = \frac{25}{60}$

$\frac{18}{60} < \frac{25}{60}$

$\frac{3}{10} < \frac{5}{12}$

$x < y < z$

22. **C** Add a few of the terms to see if there is a pattern:

$$\left(\frac{2}{2} + \frac{1}{2}\right) \cdot \left(\frac{3}{3} + \frac{1}{3}\right) \cdot \left(\frac{4}{4} + \frac{1}{4}\right) \cdot \ldots \cdot \left(\frac{22}{22} + \frac{1}{22}\right) = \frac{3}{2} \cdot \frac{4}{3} \cdot \frac{5}{4} \cdot \ldots \cdot \frac{22}{21} \cdot \frac{23}{22}$$

Notice that the numerator of the first fraction is the same as the denominator of the second and so on. Simplify all the matching numbers up to the last fraction. Simplify 3 by 3, 4 by 4, and finally 22 by 22.

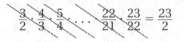

$$\frac{3}{2} \cdot \frac{4}{3} \cdot \frac{5}{4} \cdot \ldots \cdot \frac{22}{21} \cdot \frac{23}{22} = \frac{23}{2}$$

23. **D** Simplify the fractions inside the parentheses before adding.

$\frac{55 \div 11}{33 \div 11} = \frac{5}{3}$ and $\frac{5555 \div 1111}{3333 \div 1111} = \frac{5}{3}$

$\left(\frac{5}{3} + \frac{5}{3} + \frac{5}{3}\right) \cdot \frac{8}{5} = \frac{15}{3} \cdot \frac{8}{5} = 8$

24. **A** Write each fraction as a mixed number.

$m = \frac{40}{39} = 1\frac{1}{39}$ $k = \frac{400}{399} = 1\frac{1}{399}$ $n = \frac{4000}{3999} = 1\frac{1}{3999}$

Since the integer parts are equal, the one with the largest fraction is the greatest.

$m > k > n$

25. **D** Do not attempt to add the given fractions. In questions that are seemingly very tedious like this one, write out a few steps until you recognize a pattern.

If each numerator and each denominator is increased by one, we get

$\frac{16}{17} + \frac{17}{18} + \frac{18}{19} + \ldots + \frac{62}{63} + \frac{63}{64}$.

> **REMEMBER**
>
> Whenever you notice calculations that might take a very long time, look for a pattern that can simplify your work.

Since we need the difference, subtract the original from the new series of additions. Realize that most terms cancel each other.

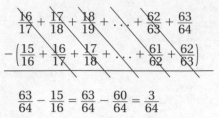

$$\frac{63}{64} - \frac{15}{16} = \frac{63}{64} - \frac{60}{64} = \frac{3}{64}$$

26. **C** After Rajiv, there were $999 - 962 = 37$ customers. Before Bianca, there were 15 customers. The total is $37 + 15 = 52$.

27. **B** If you are trying to make the largest number as large as possible, the other 4 numbers should be as small as possible. The 4 smallest 2-digit different integers are 10, 11, 12, and 13.

$$130 - (10 + 11 + 12 + 13) = 130 - 46 = 84$$

28. **D** First find the prime factorization of 24: $24 = 2 \cdot 2 \cdot 2 \cdot 3$.

$x^3 = 2 \cdot 2 \cdot 2 \cdot 3 \cdot y$

To make the right side a perfect cube, y needs to be $3 \cdot 3$. That way, the right side becomes $2 \cdot 2 \cdot 2 \cdot 3 \cdot 3 \cdot 3$, which equals $2^3 \cdot 3^3 = 6^3$. So $x = 6$ and $y = 9$, and $x + y = 15$.

29. **C** $V - 3 \overline{)W} \quad \begin{array}{c} V + 3 \\ \hline \end{array} \quad R = 5$

$W = (V - 3)(V + 3) + 5$
$W = V^2 - 9 + 5$
$W = V^2 - 4$

W has to be 4 less than the square of an integer, $45 = 49 - 4$. So W could be 45. Other answer choices do not make W an integer.

30. **E** There are 7 sections between A and H, so each section is $\frac{1}{7}$ long. AG is 6 sections long, and DH is 4 sections long. So AG is 2 sections longer that DH.
2 sections $= 2 \cdot \frac{1}{7} = \frac{2}{7}$.

SECTION 2.2: RATIOS AND PROPORTIONS

- A ratio expresses a mathematical relationship between two quantities. Specifically, the ratio is the quotient of two quantities.

   Example: If the ratio of girls to boys in one class is 5 to 4, we can express it as $\frac{5}{4}$ or $5 : 4$ or 1.25.

- A proportion is an equation where two ratios are equal. $\frac{a}{b} = \frac{c}{d}$

- The easiest way to solve a proportion with an unknown is to cross multiply.

   If $\frac{a}{b} = \frac{c}{d}$, then $b \cdot c = a \cdot d$.

- When setting up proportions, make sure the units of each side match each other.

  Example: The ratio of the length of a rectangle to its width is $9 : 7$. If the length of the rectangle is 72 inches, what is the perimeter of the rectangle?

  $$\frac{\text{length}}{\text{width}} = \frac{9}{7} \qquad \frac{9}{7} = \frac{72}{\text{width}} \qquad 9 \cdot w = 72 \cdot 7 \qquad \text{width} = \frac{72 \cdot 7}{9} = 56$$

  Perimeter $= 2(w + l) = 2(56 + 72) = 256$

Direct Proportions

- Quantities x and y are directly proportional if their ratio is constant: $\frac{y}{x} = k$ (or $y = k \cdot x$), where k is a constant. In other words, $\frac{y}{x}$ always has the same numerical value.

  Example: x and y are directly proportional. When $x = -6$, $y = 8$. What is the value of y when $x = 3$?

  Set up the equation as $y = k \cdot x$, and find k using the initial conditions.

  $y = k \cdot x$

  $k = \frac{y}{x} = \frac{8}{-6} = \frac{4}{-3}$

  $y = \frac{4}{-3} \cdot x$

  Plug in $x = 3$ to find y:

  $y = \frac{4}{-3} \cdot 3 = -4$

  Alternatively, set up a proportion since $\frac{y}{x}$ is always constant:

  $\frac{y_1}{x_1} = \frac{y_2}{x_2}$

  $\frac{8}{-6} = \frac{y_2}{3}$ Cross multiply.

  $24 = -6 \cdot y_2$

  $y_2 = -4$

> **REMEMBER**
>
> If an increase in one variable causes an increase in another variable, these two variables are directly proportional.

Inverse Proportions

- Quantities x and y are inversely proportional if their product is constant: $x \cdot y = k$, where k is a constant. In other words, $x \cdot y$ always has the same numerical value.

  Example: x and y are inversely proportional. When $x = 2$, $y = \frac{5}{2}$. What is x when $y = \frac{1}{5}$?

  Set up the equation as $x \cdot y = k$, and find k using the initial conditions.

  $x \cdot y = k$

  $k = 2 \cdot \frac{5}{2} = 5$

  $x \cdot y = 5$

  Plug in $y = \frac{1}{5}$ to find x:

  $x \cdot \frac{1}{5} = 5$

  $x = 25$

> **REMEMBER**
>
> If an increase in one variable causes a decrease in another variable, these two variables are inversely proportional.

Alternatively, set up an equation since $x \cdot y$ is always constant:

$$x_1 \cdot y_1 = x_2 \cdot y_2$$

$$2 \cdot \frac{5}{2} = x_2 \cdot \frac{1}{5}$$

$$x_2 = 25$$

SAMPLE PROBLEMS

───────────────── EXAMPLE 1 ─────────────────

If $\frac{4}{5} = \frac{n}{35}$ then $n = ?$

$n \cdot 5 = 4 \cdot 35$ Do not multiply yet. Divide both sides by 5 to simplify.

$$n = \frac{4 \cdot \overset{7}{\cancel{35}}}{\underset{1}{\cancel{5}}} = 28$$

───────────────── EXAMPLE 2 ─────────────────

If a lawyer charges \$350 for half an hour of her time, how many hours did she work if her invoice was \$2,450?

REMEMBER

Since the GMAT is a no calculator test, simplify whenever possible before you carry out the calculations.

$$\frac{\$350}{0.5\,\text{hr}} = \frac{\$2,450}{h\,\text{hr}}$$

$$\$2,450 \cdot 0.5 = \$350 \cdot h$$

$$h = \frac{\overset{7}{\cancel{\$2,450}} \cdot 0.5}{\underset{1}{\cancel{\$350}}} = 7 \cdot 0.5 = 3.5 \text{ hrs}$$

Alternatively, find her hourly rate. If she charges \$350 for half an hour, her hourly rate is \$700 per hour.

The total time can be found by dividing the total fee by the hourly fee: $\frac{\$2,450}{\$700} = 3.5$ hrs.

───────────────── EXAMPLE 3 ─────────────────

Antonio reads 100 pages in 12 minutes. How many pages can he read in one hour?

$$\frac{12\,\text{min}}{100\,\text{pages}} = \frac{60\,\text{min}}{n\,\text{pages}}$$

$$n = \frac{\overset{5}{\cancel{60}} \cdot 100}{\underset{1}{\cancel{12}}} = 500 \text{ pages}$$

EXAMPLE 4

If $m \cdot n = k \cdot t$, which of the following fractions are not equivalent?

(A) $\dfrac{m}{k} = \dfrac{t}{n}$

(B) $\dfrac{k}{m} = \dfrac{n}{t}$

(C) $\dfrac{k}{n} = \dfrac{m}{t}$

(D) $\dfrac{n}{k} = \dfrac{m}{t}$

(E) $\dfrac{t}{m} = \dfrac{n}{k}$

For each answer choice, cross multiply and see if it matches the given equation:

(A) $\dfrac{m}{k} = \dfrac{t}{n} \rightarrow m \cdot n = t \cdot k$

(B) $\dfrac{k}{m} = \dfrac{n}{t} \rightarrow t \cdot k = m \cdot n$

(C) $\dfrac{k}{n} = \dfrac{m}{t} \rightarrow t \cdot k = m \cdot n$

(D) $\dfrac{n}{k} = \dfrac{m}{t} \rightarrow t \cdot n = m \cdot k$

(E) $\dfrac{t}{m} = \dfrac{n}{k} \rightarrow t \cdot k = m \cdot n$

The answer is (D).

EXAMPLE 5

If $a : b = 3 : 5$ and $b : c = 3 : 4$, what is $a : c$?

The question can be written as follows:

$\dfrac{a}{b} = \dfrac{3}{5}$ and $\dfrac{b}{c} = \dfrac{3}{4}$ then $\dfrac{a}{c} = ?$

Solution 1:   Cross multiply both ratios.

$5a = 3b$ and $4b = 3c$ Since we are looking for a ratio of a to c, we need to eliminate b. Solve for b in the first equation.

$5a = 3b \quad \rightarrow \quad b = \dfrac{5a}{3}$ Plug this b into the second equation.

$4 \cdot \dfrac{5a}{3} = 3c$ Multiply both sides by 3.

$20a = 9c$ Divide both sides by $20c$ to obtain $\dfrac{a}{c}$.

$\dfrac{20a}{20c} = \dfrac{9c}{20c} \rightarrow \dfrac{a}{c} = \dfrac{9}{20}$

Solution 2:   Multiply the two ratios to eliminate b and get the third ratio, $\dfrac{a}{c}$.

$\dfrac{a}{b} \cdot \dfrac{b}{c} = \dfrac{a}{c}$

$\dfrac{3}{5} \cdot \dfrac{3}{4} = \dfrac{9}{20} = \dfrac{a}{c}$

1. If $\frac{a}{b} = \frac{b}{c} = 5$, what is the value of $\frac{a}{c}$?

   (A) $\frac{12}{5}$

   (B) $\frac{1}{5}$

   (C) 1

   (D) 5

   (E) 25

2. If $\frac{m}{3} = \frac{k}{5} = \frac{n}{6}$, what is the value of $\frac{m+k}{n}$?

   (A) $\frac{1}{3}$

   (B) $\frac{1}{2}$

   (C) $\frac{5}{6}$

   (D) $1\frac{1}{6}$

   (E) $1\frac{1}{3}$

3. The ratio of brown eggs to white eggs in a basket is 1.5 to 1.8. If there are fewer than 25 brown eggs in the basket, what is the highest possible number of total eggs in the basket?

   (A) 22

   (B) 24

   (C) 44

   (D) 55

   (E) 60

4. If $\frac{a}{b} = \frac{3}{5}$, then $\frac{2a+3b}{b} = $?

   (A) $4\frac{1}{5}$

   (B) 3

   (C) $2\frac{3}{5}$

   (D) $2\frac{1}{5}$

   (E) $1\frac{1}{3}$

5. The ratio of fiction books to nonfiction books in Abi's library is 4 : 5. If she buys 6 more books of each kind, the ratio would become 5 : 6. How many fiction books does Abi's library currently have?

   (A) 20

   (B) 24

   (C) 25

   (D) 30

   (E) 36

6. If $\dfrac{x}{x+y} = \dfrac{4}{3}$, what is the value of $\dfrac{x}{y}$?

   (A) $\dfrac{2}{3}$

   (B) $\dfrac{3}{4}$

   (C) $\dfrac{1}{3}$

   (D) $-\dfrac{1}{4}$

   (E) -4

7. If $\dfrac{a}{4} = \dfrac{b}{5} = \dfrac{c}{6}$ and $a + b + c = 60$, what is the value of $\dfrac{b \cdot c}{a + c}$?

   (A) 3

   (B) 6

   (C) 8

   (D) 12

   (E) 24

8. Three teenagers aged 12, 13, and 15 share $240 proportional to their ages. How much does the youngest get?

   (A) $6

   (B) $12

   (C) $40

   (D) $72

   (E) $78

9. If a machine can fill 1400 soda cans in one hour, how many can it fill in 84 minutes?

   (A) 1540

   (B) 1750

   (C) 1960

   (D) 2040

   (E) 2240

10. If $\dfrac{x}{m} = \dfrac{y}{k} = \dfrac{z}{t} = 3$, what is the value of $\dfrac{m \cdot y \cdot t}{x \cdot k \cdot z}$?

    (A) 27

    (B) 9

    (C) 3

    (D) $\dfrac{1}{3}$

    (E) $\dfrac{1}{9}$

11. The ratio of mountain bikes to road bikes in a parking lot is 3 : 5. What percent of bikes in this parking lot are road bikes?

    (A) 32.5%

    (B) 40%

    (C) 60%

    (D) 62.5%

    (E) 75%

12. If $\dfrac{a+b}{3} = \dfrac{a-b}{2}$, which of the following must be true?

(A) a is equal to b

(B) a is 5 times b

(C) a is 3 times b

(D) b is 3 times a

(E) b is 2 times a

SECTION 2.2—SOLUTIONS

1	2	3	4	5	6	7	8	9	10	11	12
E	E	C	A	B	E	D	D	C	D	D	B

1. **E** $\dfrac{a}{b} = 5$ and $\dfrac{b}{c} = 5$

To find $\dfrac{a}{c}$, multiply $\dfrac{a}{b}$ by $\dfrac{b}{c}$. The b's will cancel each other.

$$\frac{a}{b} \cdot \frac{b}{c} = \frac{5}{1} \cdot \frac{5}{1}$$

$$\frac{a}{c} = 25$$

Alternatively, cross multiply each ratio, solve for b, and substitute.

If $\dfrac{a}{b} = 5$, then $a = 5b$.   If $\dfrac{b}{c} = 5$, then $b = 5c$. Substitute $5c$ into the first equation instead of b.

$a = 5 \cdot (5c)$

$a = 25c$

$\dfrac{a}{c} = 25$

2. **E** Split the fraction.

$$\frac{m+k}{n} = \frac{m}{n} + \frac{k}{n}$$

Find $\dfrac{m}{n}$ and $\dfrac{k}{n}$ separately and then add.

If $\dfrac{m}{3} = \dfrac{n}{6}$, then $\dfrac{m}{n} = \dfrac{3}{6}$.

If $\dfrac{k}{5} = \dfrac{n}{6}$, then $\dfrac{k}{n} = \dfrac{5}{6}$.

$$\frac{m}{n} + \frac{k}{n} = \frac{3}{6} + \frac{5}{6} = \frac{8}{6} = \frac{4}{3} = 1\frac{1}{3}$$

Alternatively, let $\dfrac{m}{3} = \dfrac{k}{5} = \dfrac{n}{6} = t$, a constant number. We can represent m, k, and n in terms of t.

If $\dfrac{m}{3} = t$, then $m = 3t$. Similarly, $k = 5t$ and $n = 6t$.

$$\frac{m+k}{n} = \frac{3t+5t}{6t} = \frac{8t}{6t} = \frac{8}{6} = \frac{4}{3} = 1\frac{1}{3}$$

3. **C** The ratio is $\frac{B}{W} = \frac{1.5}{1.8} = \frac{15}{18} = \frac{5}{6}$ since the number of eggs need to be an integer. The highest multiple of 5 that is less than 25 is 20. So there can be a maximum of 20 brown eggs.

$$\frac{5}{6} = \frac{20}{W} \qquad \text{Cross multiply.}$$

$$15W = 120$$

$$W = 24$$

Total $= 20 + 24 = 44$

4. **A** Split the fraction.

$$\frac{2a + 3b}{b} = \frac{2a}{b} + \frac{3b}{b} = \frac{2a}{b} + 3$$

Since $\frac{a}{b} = \frac{3}{5}$, plug it in the equation.

$$2 \cdot \frac{3}{5} + 3 = \frac{6}{5} + \frac{3}{1} = \frac{6}{5} + \frac{15}{5} = \frac{21}{5} = 4\frac{1}{5}$$

Alternatively, let $a = 3x$ and $b = 5x$ since $\frac{a}{b} = \frac{3}{5} = \frac{3x}{5x}$

$$\frac{2a + 3b}{b} = \frac{2 \cdot 3x + 3 \cdot 5x}{5x} = \frac{6x + 15x}{5x} = \frac{21x}{5x} = \frac{21}{5} = 4\frac{1}{5}$$

5. **B** Let the number of fiction books be $4x$. The number of nonfiction books then becomes $5x$ since their ratio is $\frac{4}{5}$. This means $\frac{4}{5} = \frac{4x}{5x}$.

$$\frac{4x + 6}{5x + 6} = \frac{5}{6} \qquad \text{Cross multiply.}$$

$$6 \cdot (4x + 6) = 5 \cdot (5x + 6)$$

$$24x + 36 = 25x + 30$$

$$36 = x + 30$$

$$6 = x$$

The number of fiction books $= 4x = 4 \cdot 6 = 24$.

6. **E** Remember that $\dfrac{x}{x + y} \neq \dfrac{x}{x} + \dfrac{x}{y}$

$$\frac{x + y}{x} = \frac{3}{4} \qquad \text{Cross multiply and distribute the 4.}$$

$$4x + 4y = 3x \qquad \text{Subtract } 4x.$$

$$4y = -x \qquad \text{Divide by } -y \text{ to get } \frac{x}{y}.$$

$$\frac{x}{y} = -4$$

7. **D** Let $\frac{a}{4} = \frac{b}{5} = \frac{c}{6} = k$, a constant number. We can represent a, b, and c in terms of k.

$a = 4k$, $b = 5k$, and $c = 6k$

If $a + b + c = 60$, then $4k + 5k + 6k = 60$.

$$15k = 60$$

$$k = \frac{60}{15} = 4$$

If $k = 4$, then $a = 4k = 16$, $b = 5k = 20$, and $c = 6k = 24$.

$$\frac{b \cdot c}{a + c} = \frac{20 \cdot 24}{16 + 24} = \frac{20 \cdot 24}{40} = 12$$

8. **D** If the 12 year old gets $12x$, the 13 year old gets $13x$ and the 15 year old gets $15x$.

   The total is $12x + 13x + 15x = 40x$.

   $40x = 240$

   $x = \$6$ per year

   The youngest one gets $6 \cdot 12 = \$72$.

9. **C** Set up a proportion.

   $\dfrac{\text{bottles}}{\text{minutes}} = \dfrac{1400}{60} = \dfrac{x}{84}$ Cross multiply.

   $84 \cdot 1400 = 60 \cdot x$ Do not multiply yet. First divide both sides by 60.

   $x = \dfrac{84 \cdot 1400}{60}$ Divide by 10 and then by 6.

   $\dfrac{14 \cdot 140}{1} = 1960$

10. **D** $\dfrac{m \cdot y \cdot t}{x \cdot k \cdot z}$

    Split the fraction into the multiplication of 3 fractions, $\dfrac{m}{x} \cdot \dfrac{y}{k} \cdot \dfrac{t}{z}$.

    If $\dfrac{x}{m} = 3$, then $\dfrac{m}{x} = \dfrac{1}{3}$.   We are given $\dfrac{y}{k} = 3$.   If $\dfrac{z}{t} = 3$, then $\dfrac{t}{z} = \dfrac{1}{3}$.

    $\dfrac{m}{x} \cdot \dfrac{y}{k} \cdot \dfrac{t}{z} = \dfrac{1}{3} \cdot 3 \cdot \dfrac{1}{3} = \dfrac{1}{3}$

11. **D** Let the number of mountain bikes be $3x$. The number road bikes becomes $5x$ since their ratio is $3 : 5$. Then the total number of bikes is $3x + 5x = 8x$.

    The ratio of road bikes to total bikes $\dfrac{5x}{8x} = \dfrac{5}{8}$.

    To convert to percentages, divide 5 by 8 and multiply by 100%.

    $\dfrac{5}{8} \cdot 100\% = \dfrac{500}{8}\% = 62.5\%$

12. **B** $\dfrac{a + b}{3} = \dfrac{a - b}{2}$ Cross multiply.

    $3(a - b) = 2(a + b)$ Distribute.

    $3a - 3b = 2a + 2b$ Subtract $2a$ and add $3b$.

    $a = 5b$ This means a is 5 times b.

SECTION 2.3—PERCENTAGES

- A percentage is a way to represent a fraction with a denominator of 100.

   % (percent) means "divide by 100" or $\dfrac{}{100}$ $m\%$ means $\dfrac{m}{100}$

   $2\% = \dfrac{2}{100} = 0.02$ $27\% = \dfrac{27}{100} = 0.27$ $235\% = \dfrac{235}{100} = 2.35$

- To find a percentage of a number, multiply that number by the decimal equivalent of the percentage. When you see "of," think multiplication.

   Example: What is 44% of 25?

   $0.44 \cdot 25 = 11$ or $\dfrac{44}{\cancel{100}_{4}} \cdot \cancel{25}^{1} = \dfrac{44}{4} = 11$

Alternatively, you can set up a proportion:

$$\frac{\text{Part}}{\text{Whole}} = \frac{44}{100} = \frac{x}{25} \qquad \text{Cross multiply.}$$

$$x = \frac{44 \cdot \overset{1}{\cancel{25}}}{\underset{4}{\cancel{100}}} = \frac{44}{4} = 11$$

- To find what percent one number is of another number, simply divide the numbers and convert the result into a percentage by multiplying it by 100% because 100% = 1.

Example: 8 is what percent of 25?

$$\frac{8}{\underset{1}{\cancel{25}}} \cdot \overset{4}{\cancel{100}}\% = 32\%$$

8 is 32% of 25.

Alternatively, you can set up a proportion:

$$\frac{\text{Part}}{\text{Whole}} = \frac{x}{100} = \frac{8}{25}$$

$$x = \frac{\overset{4}{\cancel{100}} \cdot 8}{\underset{1}{\cancel{25}}} = 32$$

8 is 32% of 25.

Percentage Increase or Decrease

$$\% \text{ increase or decrease} = \frac{\text{New} - \text{Original}}{\text{Original}} \cdot 100\%$$

Example: A supermodel slims down from 120 lbs to 105 lbs for a show. What is the percentage decrease in her weight?

$$\frac{\text{New} - \text{Original}}{\text{Original}} \cdot 100\% = \frac{105 - 120}{120} \cdot 100\% = \frac{-15}{120} \cdot 100\% = -12.5\%$$

Note that the negative sign means a decrease.

Multiple Percentage Changes

- DO NOT add or subtract percentage changes. Calculate each separately.
- Use 100 if the question does not give you a number to start with.

Example: The population of Smallville increased by 20% from 2000 to 2001 and by 30% from 2001 to 2002. By what percent did the population increase from 2000 to 2002?

Assume the population in 2000 was 100.

The population increase from 2000 to 2001 is $100 \cdot 0.2 = 20$. So in 2001, the population was 120.

From 2001 to 2002, the increase is based on 120. So, the increase was $120 \cdot 0.3 = 36$. The population in 2002 was $120 + 36 = 156$. Since we started at 100, the last two digits of 1<u>56</u> give us the overall percentage increase: 56%. Note that this increase is greater than 20% + 30%.

<u>2000</u>	<u>2001</u>	<u>2002</u>
100	$100 \cdot 1.2 = 120$	$120 \cdot 1.3 = 156$
	+20%	+30%

SAMPLE PROBLEMS

EXAMPLE 1

The price of a handbag is $204 after a discount of 15%. If Aisha bought the handbag at a 20% discount off the original price with her club card, how much more did she save compared with the regular sale?

(A) $48
(B) $36
(C) $24
(D) $12
(E) $6

We can set up a simple proportion to find the original price. First, calculate what percent of the original price is $204. 100% − 15% = 85%. Therefore, $204 is 85% of the original price.

$\dfrac{\text{Part}}{\text{Whole}} = \dfrac{85}{100} = \dfrac{204}{x}$ Cross multiply.

$204 \cdot 100 = 85 \cdot x$ Divide both sides by 85.

$x = \dfrac{204 \cdot 100}{85} = 240$ Simplify and divide. $240 is the original price.

The regular saving is $240 minus $204, which equals $36.

If the original price is $240, the savings with a 20% discount could be found in one of two ways.

$\dfrac{20}{100} = \dfrac{x}{240}$ $x = \dfrac{20 \cdot 240}{100} = \$48$ or $0.2 \cdot 240 = \$48$

Aisha saved $12 (= $48 − $36) more with her club card, answer (D).

EXAMPLE 2

Yao's investments gained 20% after one year and lost 15% the next year. If he made $12,000 profit overall for the two years, how much was his initial investment?

(A) $1,200,000

(B) $900,000

(C) $800,000

(D) $600,000

(E) $300,000

Since the initial investment is not given, assume it is 100 units. After one year, it increased to 120 (100 · 1.2) units. In the second year, the loss of 15% is applied to 120 units. So Yao lost $0.15 \cdot 120 = 18$ units.

Overall, the investment went from 100 units to 102 units (120 − 18). Therefore, the overall increase is 2%.

Then set up a proportion.

$$\frac{\text{Part}}{\text{Whole}} = \frac{2}{100} = \frac{\$12,000}{x}$$

$$x = \frac{\$12,000 \cdot 100}{2} = \$600,000$$

The answer is (D).

EXAMPLE 3

60% of 30% of a number is equal to 125% of 180. What is 200% of the number?

(A) 650

(B) 1,250

(C) 1,300

(D) 2,500

(E) 2,600

Translate the question into an equation.

Replace 60% by 0.6, 30% by 0.3, and 125% by 1.25. Replace "of" with multiplication.

$0.6 \cdot 0.3 \cdot x = 1.25 \cdot 180$ Divide both sides by 0.6 · 0.3.

$x = \dfrac{1.25 \cdot 180}{0.6 \cdot 0.3}$ Multiply by $\dfrac{100}{100}$ to convert all decimals into integers.

$x = \dfrac{125 \cdot 180}{6 \cdot 3} = \dfrac{125 \cdot 10}{1} = 1,250$ Simplify.

200% of $1,250 = \dfrac{200}{100} \cdot 1,250 = 2 \cdot 1,250 = 2,500$

The answer is (D).

1. What is the original price of a laptop computer that is on sale for $435 after a 13% discount?

   (A) $448
   (B) $461
   (C) $500
   (D) $513
   (E) $530

2. Manju finishes 10% of her assignment on the first day. What percent of the remaining portion does she have to finish on the second day so that only 45% of the entire assignment remains to be done?

   (A) 40%
   (B) 45%
   (C) 50%
   (D) 55%
   (E) 60%

3. A grocer buys apples for 63¢ per pound. If 10% of the apples goes bad and he still wants to make a 20% profit over his purchase price, what should be the sales price?

   (A) 66¢
   (B) 70¢
   (C) 75¢
   (D) 77¢
   (E) 84¢

4. Andre drives at a constant speed of 64 miles per hour for 4 hours and covers 80% of the total distance he is traveling. What is the remaining distance?

   (A) 64 miles
   (B) 128 miles
   (C) 256 miles
   (D) 320 miles
   (E) 512 miles

5. Kumi is taking a test that has 60 multiple-choice questions. She answers 60% of the first 20 questions correctly. If she wants to score 70% on this test, what percent of the remaining questions does she need to answer correctly?

   (A) 80%
   (B) 75%
   (C) 70%
   (D) 65%
   (E) 60%

6. What is 40% of $\frac{5}{4}$?

    (A) $\frac{5}{2}$

    (B) $\frac{5}{8}$

    (C) $\frac{3}{4}$

    (D) $\frac{1}{4}$

    (E) $\frac{1}{2}$

7. A beauty products company sells hair extensions. They increased the price of extensions by 20% from year 1 to year 2, and their sales numbers decreased by 20%. By what percent did the yearly earnings change? (earnings = price · quantity)

    (A) Increased by 4%

    (B) Increased by 2%

    (C) No change

    (D) Decreased by 2%

    (E) Decreased by 4%

8. A store receives a shipment of the latest Shoot 'Em Up video game. 70% of the games is sold at 30% profit, and the remaining 30% is sold at a loss of 20%. What is the store's percentage gain or loss from the sale of this video game?

    (A) 10%

    (B) 15%

    (C) 20%

    (D) 25%

    (E) 30%

9. 16% of a number is 1.12. What is 125% of that number?

    (A) 8.75

    (B) 7.50

    (C) 7.00

    (D) 6.75

    (E) 5.25

10. The length of a rectangle is increased by 40%. By what percent does the width need to be changed so that the area decreases by 2%?

    (A) Increased by 2%

    (B) Decreased by 42%

    (C) Decreased by 40%

    (D) Decreased by 38%

    (E) Decreased by 30%

11. Lea starts working on her homework and finishes 80% of the work in one hour. Since she is getting tired, she finishes 80% of the remaining homework in the next hour. What percent of the original homework remains unfinished after two hours?

(A) 16%

(B) 8%

(C) 4%

(D) 2%

(E) 0%

12. A certain hand sanitizer claims to kill 90% of bacteria with each use. If Sue applies the sanitizer twice in a row, what percent of the original bacteria will remain on her hands?

(A) 1%

(B) 8.1%

(C) 9%

(D) 10%

(E) 11%

SECTION 2.3—SOLUTIONS

1	2	3	4	5	6	7	8	9	10	11	12
C	C	D	A	B	E	E	B	A	E	C	A

1. **C** If it is on sale for 13%, its value is 87% of the original, $100\% - 13\% = 87\%$.

   $435 is 87% of the original price. Setting up a proportion is an easy way to find the original price.

   $\dfrac{\text{Part}}{\text{Whole}} = \dfrac{87}{100} = \dfrac{435}{x}$ Cross multiply.

   $87 \cdot x = 435 \cdot 100$ Solve for x by dividing by 87 and simplifying.

   $x = \dfrac{435 \cdot 100}{87} = \$500$

   Alternatively, set up an algebraic equation.

   $0.87 \cdot n = 435$ $n = \dfrac{435}{0.87} = \dfrac{43{,}500}{87} = 500$

2. **C** Let the entire assignment be 100 units. She finishes 10% on the first day, so 90 units remain. She wants to leave only 45% (45 units) undone after the second day. So she has to finish $90 - 45 = 45$ units on the second day. She has to finish $\dfrac{45}{90} = \dfrac{1}{2} = 50\%$ of the remaining assignment.

3. **D** Since 10% of the apples went bad, he ended up paying 63¢ for 0.9 lb. He purchased the sellable apples for $\dfrac{63¢}{0.9\text{ lb}} = 70¢$ per pound.

   If he wants to make a 20% profit ($70 \cdot 0.2 = 14¢$), the price should be $63¢ + 14¢ = 77¢$.

   Alternatively, since he wants to make 20% profit, he needs to make $63¢ \cdot 20\% = 12.6¢$ profit per pound purchased. Note that he only has 0.9 pounds left to sell after 10% went bad. His profit per pound needs to be $\dfrac{12.6¢}{0.9} = 14¢$. The sales price is $63¢ + 14¢ = 77¢$ per pound.

4. **A** distance = rate · time

He traveled $64 \cdot 4 = 256$ miles so far. The total distance can be found as follows.

$$\frac{80}{100} = \frac{256}{x} \qquad x = \frac{256 \cdot 100}{80} = 320 \text{ miles}$$

If the total distance is 320 miles, $320 - 256 = 64$ miles remain.

Alternatively, solve for the total time required since the speed is constant.

$$\frac{80}{100} = \frac{4}{x}$$

$$x = \frac{4 \cdot 100}{80} = 5 \text{ hours}$$

5 hours is the total trip. $5 - 4 = 1$ hour remains. Since distance = rate · time, the distance remaining is $1 \cdot 64 = 64$ miles.

5. **B** 60% of $20 = 0.6 \cdot 20 = 12$

She already answered 12 correct. She wants 70% of 60 questions correct. Kumi's goal is $0.7 \cdot 60 = 42$ correct.

She needs $42 - 12 = 30$ more correct answers out of 40.

$$\frac{30}{40} \cdot 100\% = 75\%$$

Alternatively, you can set up a weighted average equation. (See Chapter 4 "Word Problems.")

$$60\% \cdot 20 + x \cdot 40 = 70\% \cdot 60$$

$$0.6 \cdot 20 + 40x = 0.7 \cdot 60$$

$$12 + 40x = 42$$

$$40x = 30$$

$$x = 75\%$$

6. **E** To translate, replace 40% by either 0.4 or $\frac{40}{100}$ and replace "of" by using multiplication.

$$40\% \text{ of } \frac{5}{4} = \frac{40}{100} \cdot \frac{5}{4} = \frac{4}{10} \cdot \frac{5}{4} = \frac{1}{2}$$

7. **E** Since no numbers are given, let's assume the original price was $10 and the company sold 10 units in year 1.

Total earnings equal $\$10 \cdot 10 = \$100$ in year 1.

The new price equals $10 + (0.2 \cdot 10) = \$12$.

The new sales equal $10 - (0.2 \cdot 10) = 8$ units.

Total earnings equal $\$12 \cdot 8 = \$96$ in year 2.

$$\% \text{ change} = \frac{96 - 100}{100} \cdot 100\% = -4\% = 4\% \text{ decrease}$$

8. **B** Assume 10 games are delivered at a cost of $10 each ($100 total purchase price). So 7 of them (70%) are sold at $13 (30% profit), and 3 (30%) of them are sold at $8 (20% loss.)

Total earnings equal $7 \cdot 13 + 3 \cdot 8 = 91 + 24 = \$115$. Since the total purchase price is $100, the total percentage gain is 15%.

Alternatively, use $0.7 \cdot 0.3 - 0.3 \cdot 0.2 = 0.21 - 0.06 = 0.15 = 15\%$ gain.

9. **A** Set up a proportion for the first part.

$$\frac{16}{100} = \frac{1.12}{x}$$ Cross multiply.

$$x = \frac{112}{16} = 7$$

The number is 7.

125% of 7 = 1.25 · 7 = 8.75

10. **E** Assume the initial sides are $L = 10$ and $W = 10$. So the initial area = 100.

The length is increased by 40%, therefore the new $L = 10 · 1.4 = 14$.

The new area is 2% less, therefore the new $A = 0.98 · 100 = 98$.

New Area = New L · New W

$$98 = 14 · \text{New } W$$

$$\text{New } W = \frac{98}{14} = 7$$

Since the original width was 10, the percent decrease is $\frac{(10 - 7)}{10} = 30\%$.

11. **C** Let the amount of homework be 100 units. She finishes 80 units in the first hour, so 20 units remain after the first hour.

She finishes 80% of 20 units in the second hour. So 20 · 0.8 = 16 units are finished in the second hour.

The remaining work equals 100 − 80 − 16 = 4 units

So the remaining work is $\frac{4}{100} = 4\%$.

12. **A** Let the number of bacteria be 100 at the beginning. Since the hand sanitizer kills 90%, 10 bacteria will be left after the first application.

The second application will kill 10 · 90% = 9 more, and only 1 bacterium will be left.

The answer is 1%.

SECTION 2.4: DESCRIPTIVE STATISTICS
Average (Arithmetic Mean)

REMEMBER

To divide by 5 quickly, divide the number by 10 and multiply by 2.

- The average (or arithmetic mean) is the sum of a set of values divided by the number of values in the set.

$$\text{Average} = \frac{\text{Sum of all terms}}{\text{Number of terms}}$$

Example: The average of 8, 12, 14, 22, and 40 is $\frac{8 + 12 + 14 + 22 + 40}{5} = \frac{96}{5} = 19.2$.

- Word problems usually involve either combining or separating two groups. In these cases, start by calculating the sum of the quantity for each group since group averages cannot be added or subtracted.

Sum = Average · Number of terms

Example: The average grade of all girls in a Spanish class is 86, and the average grade of all boys is 78. If there are 12 girls and 8 boys in class, what is the class average?

The sum of all girls' grades = 86 · 12 = 1,032.

The sum of all boys' grades = 78 · 8 = 624.

The sum of all grades in the class = $1,032 + 624 = 1,656$.

The average grade = $\dfrac{\text{The sum of all grades in the class}}{\text{Total number of students}} = \dfrac{1,656}{20} = 82.8$.

- If a number n is added to all the values in a set of numbers, the average of the numbers increases by n.

    Example: The average of 4, 7, 12, and 15 is $\dfrac{4 + 7 + 12 + 15}{4} = \dfrac{38}{4} = 9.5$.
    If 3 is added to each number in the set, the average becomes
    $\dfrac{7 + 10 + 15 + 18}{4} = \dfrac{50}{4} = 12.5$, which is 3 more than 9.5.

Median

- In a list of values arranged in increasing (or decreasing) order, the median is the middle value.

    Example: The median of the set {3, 3, 4, 5, 5, 5, 7, 8, 8, 9, 12} is the 6th number, which is 5.

- If the list has an even number of values, the median is the average of the two middle values.

    Example: The median of the set {5, 5, 5, <u>6, 8</u>, 8, 9, 12} is the average of the two middle numbers, $\dfrac{(6 + 8)}{2} = 7$.

    Note that the median may or may not be one of the elements of the original set of numbers.

Mode

- In a list of values, the mode is the value that appears most frequently.

    Example: The mode of the set {3, 3, 4, 5, 5, 5, 6, 8, 8, 9, 12} is 5.

Range

- Range is the simplest measurement of spread (dispersion) of data. It is the difference between the greatest value and the least value.

    Example: The range of the set {−7, −5, 0.25, 4, 12, 14, 21} is $(21 − (−7)) = 28$.

- If a number n is added to all the values in a set of numbers, the range does not change.

    Example: The range of −4, 7, 12, and 21 is $21 − (−4) = 25$.
    If 3 is added to each number in the set, the set becomes
    −1, 10, 15, 24 and the range is $24 − (−1) = 25$.

Standard Deviation

- Standard deviation is a commonly used measure of the spread (dispersion) of data. The higher the standard deviation, the more spread out the data.
- Standard deviation shows how much variation there is from the average. The further the data are spread away from the mean, the higher the standard deviation.
- Standard deviation is also used to indicate the difference between a number and the mean.

    Example: In a set of real numbers, the mean is 7 and the standard deviation is 1.3.
    Therefore, 9.6 is 2 standard deviations more than the mean ($7 + 2 \cdot 1.3 = 9.6$).
    In addition, 3.1 is 3 standard deviations less than the mean ($7 − 3 \cdot 1.3 = 3.1$).

- Understanding and interpreting the standard deviation is more important than calculating it by hand since it is unlikely that you will be asked to calculate it.
- Two data sets with the same mean may have very different standard deviations.

   Example: $A = \{1, 3, 5, 7, 9\}$ and $B = \{3, 4, 5, 6, 7\}$

   Both sets have a mean of 5. However, set A has a higher standard deviation since its elements are more dispersed around (far away from) 5.

- To calculate the standard deviation of a set of numbers, first find the average of the set. Second, find the difference of each number from the mean and take the square of each difference. Then add the squares of each difference. Divide that sum by the number of data points. Finally, take the square root.

   Example: Both $A = \{1, 3, 5, 7, 9\}$ and $B = \{3, 4, 5, 6, 7\}$ have a mean of 5.

   1: $(1 - 5)^2 = 16$
   3: $(3 - 5)^2 = 4$
   5: $(5 - 5)^2 = 0$
   7: $(7 - 5)^2 = 4$
   9: $(9 - 5)^2 = 16$ Standard deviation $= \sqrt{\dfrac{(16 + 4 + 0 + 4 + 16)}{(5)}} \approx 2.83$

   3: $(3 - 5)^2 = 4$
   4: $(4 - 5)^2 = 1$
   5: $(5 - 5)^2 = 0$
   6: $(6 - 5)^2 = 1$
   7: $(7 - 5)^2 = 4$ Standard deviation $= \sqrt{\dfrac{(4 + 1 + 0 + 1 + 4)}{(5)}} \approx 1.41$

   As you can see, set B has a much lower standard deviation than set A.

SAMPLE PROBLEMS

EXAMPLE 1

Which of the following set of numbers has a higher mean than median?

(A) {3, 4, 5, 6, 7}

(B) {1, 3, 5, 7, 9}

(C) {1, 1, 5, 6, 7}

(D) {3, 4, 5, 7, 8}

(E) {2, 3, 5, 6, 7}

(A) {3, 4, 5, 6, 7} mean = median = 5 since the numbers are consecutive, the middle number is also the mean.

(B) {1, 3, 5, 7, 9} mean = median = 5 since the numbers are consecutive odd, the middle number is also the mean.

(C) {1, 1, 5, 6, 7} median = 5 but the mean is less than 5 since the sum of this set is less than that of the sets above or think of it being "lighter" on the left of 5. The mean will slide toward the left (smaller) since 1 is further away from 5 compared to 6 or 7.

(D) {3, 4, 5, 7, 8} median = 5 and the mean is slightly higher since this time it slides toward the right. The numbers on the right are further away from 5.

(E) {2, 3, 5, 6, 7} same as C.

The answer is (D).

EXAMPLE 2

5 is added to each of the numbers in a set of 12 integers. Which of the following must be true?

(A) The median stays the same.

(B) The range increases by 5.

(C) The mean stays the same.

(D) The standard deviation does not change.

(E) The mode does not change.

(A) False. The median increases by 5 since each number increases by 5.

(B) False. The range stays the same. Both the largest and the smallest numbers increase by 5.

(C) False. The value of the mean increases by 5 since each number increases by 5.

(D) True. The spread of the numbers stays the same.

(E) False. The mode increases by 5 since each number increases by 5.

The answer is (D).

EXAMPLE 3

The annual average salary of 50 employees is $68,000. If 40 of these employees have an average salary of $65,000, what is the average salary of the remaining 10 employees?

(A) $60,000

(B) $68,000

(C) $74,000

(D) $78,000

(E) $80,000

Average	·	Number	=	Total
68,000		50 employees		3,400,000
65,000		40 employees		2,600,000
x		10 employees		800,000

$3,400,000 - 2,600,000 = 800,000$

The remaining 10 employees have a total salary of $800,000.

Their average salary is $\dfrac{\$800,000}{10} = \$80,000$

The answer is (E).

Alternatively, set up a weighted average equation.

$$\frac{40 \cdot 65,000 + 10 \cdot x}{50} = 68,000$$

$$2,600,000 + 10x = 3,400,000$$

$$10x = 800,000$$

$$x = 80,000$$

1. The average of 16 numbers is $\frac{15}{4}$. If 8 more numbers with a sum of 132 are added, what is the average of the 24 numbers?

   (A) $\frac{15}{8}$

   (B) $\frac{23}{8}$

   (C) $\frac{93}{8}$

   (D) 8

   (E) 30

2. 13 people have an average age of 27. What will their average age be in 3 years?

   (A) 30

   (B) 33

   (C) 39

   (D) 40

   (E) 43

3. Roberto spent a total of 12 days studying for the GMAT math section. On each of the first 4 days, he solved 80 questions. During the next 5 days, he solved 50 questions every day. For the last 3 days, he solved a total of 54 questions. How many questions did he solve per day on average?

   (A) 48.3

   (B) 52

   (C) 61.3

   (D) 62

   (E) 65

4. The average (arithmetic mean) of a certain normal distribution is 22, and the standard deviation is 4. What is the difference between the value that is 2.5 standard deviations less than the mean and the value that is 2.5 standard deviations more than the mean?

   (A) 4

   (B) 8

   (C) 10

   (D) 20

   (E) 30

5. The average of a, b, and 12 is 42. What is the average of a, b, 4, and 50?

   (A) 42

   (B) 36

   (C) 32

   (D) 24

   (E) 17.5

6. There are 21 students and a driver on a bus. Six of the students are 13 years old, seven of them are 14 years old, and the rest are 15 years old. If the driver is 34 years old, the mean age of everyone on the bus is how much more than the median age?

    (A) 0 year
    (B) 1 year
    (C) 7 years
    (D) 14 years
    (E) 15 years

7. The average of six numbers is 15.5. If another number is added to the list, the average becomes 10. What is the new number?

    (A) 5.5
    (B) 0
    (C) −5.5
    (D) −22.5
    (E) −23

8. On Tuesday, Manu's Bookstore made a profit of 20% on 60% of the items sold and lost 10% on 30% of the items sold. If all books have the same sticker price and the bookstore had no profit or loss on the remaining items, what was the average profit on Tuesday?

    (A) 5.0% profit
    (B) 7.5% profit
    (C) 9.0% profit
    (D) 10.0% profit
    (E) 12.5% profit

9. A printing shop has 10 printers working at a rate of 2,500 pages per printer per day on average. The management is considering two different new high-speed printers. Printer A has a capacity of 5,500 pages per day. Printer B has a capacity of 8,500 pages per day. How much more would their daily printing average be if they invest in two printer Bs versus two printer As?

    (A) 100 pages per printer per day
    (B) 200 pages per printer per day
    (C) 300 pages per printer per day
    (D) 400 pages per printer per day
    (E) 500 pages per printer per day

10. 6, −1, 1, 2, −3, 3, −3, 4, −2, 0, 5

    Which of the following must be true for the list of numbers above?

    (A) mode < mean < median
    (B) mean < mode < median
    (C) median < mean < mode
    (D) mode < median < mean
    (E) median < mode < mean

11. The average of x and y is k. The average of m, n, and p is t. What is the average of x, y, m, n, and p in terms of k and t?

(A) $\dfrac{(k+t)}{2}$

(B) $\dfrac{(k+t)}{5}$

(C) $\dfrac{(3k+2t)}{5}$

(D) $\dfrac{(2k+3t)}{5}$

(E) $\dfrac{5(k+t)}{6}$

12. $P, B, 2, 4, -2, 8, -1, 0$

P and B are prime numbers, and

$P + B = 12$. What is the median of the list of numbers above?

(A) 1

(B) 2

(C) 3

(D) 4

(E) 5

SECTION 2.4—SOLUTIONS

1	2	3	4	5	6	7	8	9	10	11	12
D	A	B	D	A	B	E	C	E	D	D	C

1. **D** The sum of the first 16 numbers is

Number of terms · Average $= 16 \cdot \dfrac{15}{4} = 60$

After adding the new numbers, the new total is $60 + 132 = 192$.

Since we now have 24 numbers, the new average is $\dfrac{192}{24} = 8$.

2. **A** Since each person will be 3 years older, the average will increase by 3 and become 30.

Alternatively, do the following calculations.

Sum $= 13 \cdot 27 = 351$

Increase $= 13 \cdot 3 = 39$

The new sum $= 351 + 39 = 390$

There are still 13 people.

Average $= \dfrac{390}{13} = 30$

3. **B** First find the total number of questions solved.

$(4 \cdot 80) + (5 \cdot 50) + 54 = 320 + 250 + 54 = 624$

The average is the $\dfrac{\text{total number of questions}}{\text{number of days}} = \dfrac{624}{12} = 52$.

4. **D** The value that is 2.5 standard deviations less than the mean is $22 - (2.5 \cdot 4) = 12$.

The value that is 2.5 standard deviations more than the mean is $22 + (2.5 \cdot 4) = 32$.

$32 - 12 = 20$

5. **A** The average of a, b, and 12 means adding the three terms and then dividing by 3.

$$\frac{a + b + 12}{3} = 42$$
$$a + b + 12 = 126$$
$$a + b = 114$$

The average of a, b, 4, and 50 means adding the four terms and then dividing by 4.

$$\frac{a + b + 4 + 50}{4} = \frac{114 + 54}{4} = \frac{168}{4} = 42$$

6. **B** The total age can be found as:

$(6 \cdot 13) + (7 \cdot 14) + (8 \cdot 15) + 34 = 330$

The average is $\frac{330}{22} = 15$.

Since 22 people are on the bus, the median age will be the average of the ages of the 11th and 12th persons once we list the ages from least to greatest. Both of those people are 14 years old. So the median is 14.

Mean $-$ Median $= 15 - 14 = 1$

7. **E** The original sum $=$ Average $\cdot$ Number of terms $= 6 \cdot 15.5 = 93$.

The sum after the new number is added $=$ Average $\cdot$ Number of terms $= 7 \cdot 10 = 70$.

The number must be $70 - 93 = -23$.

8. **C** Let the total number of items be 10 and each book have a sticker price of $10. The total value is $10 \cdot 10 = 100$.

6 items were sold at 20% profit: Profit $= 6 \cdot \$10 \cdot 0.2 = \$12$

3 items were sold at 10% loss: Loss $- 3 \cdot \$10 \cdot 0.1 = \$3$

Total profit on 10 books $= 12 - 3 = \$9$

Percent profit $= \dfrac{\$9}{\$100} = 9\%$

Alternatively, use the following equation.

$0.6 \cdot 0.2 - 0.3 \cdot 0.1 = 0.09 = 9\%$

9. **E** Option 1: Buy 2 printer As:

$$\text{Average} = \frac{\text{Total pages}}{\text{\# of printers}} = \frac{(10 \cdot 2{,}500) + (2 \cdot 5{,}500)}{12} = \frac{25{,}000 + 11{,}000}{12} = 3{,}000$$

Option 2: Buy 2 printer Bs:

$$\frac{(10 \cdot 2{,}500) + (2 \cdot 8{,}500)}{12} = \frac{25{,}000 + 17{,}000}{12} = 3{,}500$$

The difference is $3{,}500 - 3{,}000 = 500$ pages per printer per day.

Alternatively, look at the increase in capacity only.

Option 1: $\dfrac{2 \cdot 5{,}500}{12} = \dfrac{11{,}000}{12}$

Option 2: $\dfrac{2 \cdot 8{,}500}{12} = \dfrac{17{,}000}{12}$

The difference $= \dfrac{17{,}000}{12} - \dfrac{11{,}000}{12} = \dfrac{6{,}000}{12} = 500$

10. **D** First list the numbers from least to greatest.

$-3, -3, -2, -1, 0, 1, 2, 3, 4, 5, 6$

The median is the middle number, which is 1.

To find the mean, add all the numbers. Start with adding -1 and 1, -2 and 2, and so forth. The sum is 12.

The mean $= \dfrac{12}{11} = 1\dfrac{1}{11}$.

The mode is -3 since it repeats twice.

mode $<$ median $<$ mean

11. **D** The average of x and y means add x and y and then divide by 2.

$\dfrac{x + y}{2} = k \quad \rightarrow \quad x + y = 2k$

The average of m, n, and p means add m, n, and p and then divide by 3.

$\dfrac{m + n + p}{3} = t \quad \rightarrow \quad m + n + p = 3t$

Average of all $= \dfrac{(x + y) + (m + n + p)}{5} = \dfrac{2k + 3t}{5}$

12. **C** If $P + B = 12$ and if P and B are prime numbers, the only possible pair is 5 and 7.

List all numbers from least to greatest. $\qquad -2, -1, 0, 2, 4, 5, 7, 8$

Since there are an even number of terms, two numbers are in the middle. To find the median, average the two middle terms.

The median is $\dfrac{(2 + 4)}{2} = 3$.

SECTION 2.5: COUNTING METHODS AND PROBABILITY
Fundamental Principle of Counting

- If one event can happen in m different ways and another event can happen in n different ways, the two events together (m <u>and</u> n) can happen in $m \cdot n$ different ways.

  Example: If you can choose 1 pants, 1 sweater, and 1 pair of shoes from 4 pants, 5 sweaters, and 3 pairs of shoes, how many different outfits can you create?

  $4 \cdot 5 \cdot 3 = 60$

- If one event can happen in m different ways and another event can happen in n different ways, either m or n can happen in ($m + n$) different ways.

  Example: If you can choose either 1 sandwich or 1 soup from a list of 4 sandwiches and 5 soups, how many different options do you have?

  $4 + 5 = 9$

Factorial

- If n is an integer greater than 1, n factorial ($n!$) is defined as the product of all the integers from 1 to n.

  $3! = 3 \cdot 2 \cdot 1$

  $6! = 6 \cdot 5 \cdot 4 \cdot 3 \cdot 2 \cdot 1$

  $n! = (n) \cdot (n - 1) \cdot (n - 2) \cdot \ldots \cdot 2 \cdot 1$

- By definition, $1! = 1$ and $0! = 1$.

Ordering

- The number of ways a set of objects can be ordered is found by $n!$

  Example: How many different ways can 5 different books be ordered on a shelf?

  $5! = 5 \cdot 4 \cdot 3 \cdot 2 \cdot 1 = 120$

  Example: In how many ways can 4 people sit on a bench?

  $4! = 4 \cdot 3 \cdot 2 \cdot 1 = 24$

Permutation

- A permutation is an arrangement of a group of objects in a particular order.
- If the order in which the members are chosen creates distinct arrangements, such counting problems involve permutations.

  Example: At school, 4 candidates are running for 3 different positions, president, vice president, and secretary. How many different outcomes are possible?

  Ask the following questions:

  1. In how many different ways can I pick the president among 4 students?   4
  2. Once I pick the president, in how many different ways can I pick the vice president?   3
  3. Once I pick the president and vice president, in how many different ways can I pick the secretary?   2

  Multiply these numbers together because a president, <u>and</u> a vice president, <u>and</u> a secretary must be chosen:

  $4 \cdot 3 \cdot 2 = 24$

- The following formula can be used to find the number of possible permutations of k objects chosen from a set of n different objects when $n \geq k$.

  $P(n, k) = \dfrac{n!}{(n - k)!}$

  In the previous example, the number of possible outcomes are

  $P(4, 3) = \dfrac{4!}{(4 - 3)!} = \dfrac{4 \cdot 3 \cdot 2 \cdot 1}{1} = 24.$

  Example: If 6 athletes are running in a race, in how many ways can the gold, silver, and bronze medals be distributed?

  In this case, order clearly matters. Use permutation.

  $P(6, 3) = \dfrac{6!}{(6 - 3)!} = \dfrac{6 \cdot 5 \cdot 4 \cdot 3 \cdot 2 \cdot 1}{3 \cdot 2 \cdot 1} = 6 \cdot 5 \cdot 4 = 120$

Combination

- A combination is grouping of objects in which the order or arrangement of the objects is irrelevant. For example, if you are picking four toppings for your pizza, the order in which you pick them is not important.
- If the order in which the members are chosen makes no difference, such counting problems involve combinations.

Example: How many different groups of 3 representatives can be chosen from a group of 5 students?

Let the students be A, B, C, D, and E. A set of representatives can be selected as follows.

ABC	ACD	BCD	CDE
ABD	ACE	BCE	
ABE	ADE	BDE	

There are 10 combinations.

- The following formula can be used to find the number of possible combinations of k objects chosen from a set of n different objects when $n \geq k$.

$$C(n, k) = \frac{n!}{(n - k)! \cdot k!}$$

In the previous example, the number of groups are

$$C(5, 3) = \frac{5!}{(5 - 3)! \cdot 3!} = \frac{5 \cdot 4 \cdot 3 \cdot 2 \cdot 1}{(2 \cdot 1) \cdot (3 \cdot 2 \cdot 1)} = 10$$

Example: In a certain test, students are allowed to pick any 4 questions out of 6 questions provided. In how many different ways can Kim select her questions?

In this case, the order she picks her questions does not matter. Use combination.

$$C(6, 4) = \frac{6!}{(6 - 4)! \cdot 4!} = \frac{6 \cdot 5 \cdot 4 \cdot 3 \cdot 2 \cdot 1}{(2 \cdot 1) \cdot (4 \cdot 3 \cdot 2 \cdot 1)} = 3 \cdot 5 = 15$$

Discrete Probability

- Probability is the chance or likelihood that a certain event will happen.
- The probability of an event happening is the ratio of the number of ways the event can happen to the total number of possible outcomes.

$$P(E) = \frac{\text{Number of outcomes giving the desired result}}{\text{Number of all possible outcomes}}$$

- The probability of an event is a number between 0 and 1 inclusive, $0 \leq P(E) \leq 1$.

Example: What is the probability of rolling a 4 in one throw of a die?

A regular die has 6 faces, so there are 6 possible outcomes.
We only want the number 4, so the number of desired results is 1.
$$P(E) = \frac{1}{6}$$

Example: The probability of picking a blue marble from a bag that contains 7 blue marbles and 18 red marbles is $P(B) = \frac{7}{25}$.

- The probability of an event not happening is $P(\text{not } E) = 1 - P(E)$.

   Example: The probability of not picking a blue marble from a bag containing 7 blue marbles and 18 red marbles is $P(\text{not } B) = 1 - \frac{7}{25} = \frac{18}{25}$.

- $P(E) = 1$ represents a certain event.

   Example: When a fair die marked from 1 to 6 is rolled, the probability of a number less than 7 showing up is 1. It is a certain event.

- $P(E) = 0$ represents an impossible event.

   Example: When a fair die marked from 1 to 6 is rolled, the probability of an 8 showing up is 0. It is an impossible event.

   Example: In a certain car lot are only cars, trucks and SUVs. If a vehicle is selected randomly, the probability that it will be a car is $\frac{2}{7}$ and the probability that it will be a truck is $\frac{5}{14}$. If a total of 98 vehicles are in the parking lot, how many of them are SUVs?

> **REMEMBER**
>
> If you have only three colors of marbles in a bag, the probabilities of picking each of them separately will add up to 1.
>
> For example, if there are 4 blue, 5 green, and 7 white marbles in a bag, $P(B) + P(G) + P(W) = 1$
> $$\frac{4}{16} + \frac{5}{16} + \frac{7}{16} = \frac{16}{16} = 1$$

The probabilities of selecting a car, a truck and an SUV add up to 1: $P(C) + P(T) + P(S) = 1$.

$\frac{2}{7} + \frac{5}{14} + P(S) = 1$ Add the fractions.

$\frac{4}{14} + \frac{5}{14} + P(S) = 1$

$\frac{9}{14} + P(S) = 1$ Subtract $\frac{9}{14}$ from each side of the equation.

$P(S) = 1 - \frac{9}{14} = \frac{5}{14}$

If the probability of picking an SUV is $\frac{5}{14}$, you can set up a proportion.

$\frac{5}{14} = \frac{x}{98}$ Cross multiply.

$x = 98 \cdot \frac{5}{14} = 35$

There are 35 SUVs in the parking lot.

Probabilities of Two or More Events

INDEPENDENT EVENTS

- If the occurrence of event E does not affect the probability that event F occurs, E and F are independent events. The probability of both E and F happening is the product of their individual probabilities.

$P(E \text{ and } F) = P(E) \cdot P(F)$

   Example: A scarf is picked randomly from a drawer full of scarves and replaced. There are 30 scarves in the drawer. 15 are square shaped, 7 are rectangular and 8 are triangular. What is the probability of picking two triangular scarves in succession?

Since the scarves are replaced after picking, the probabilities of each pick are independent from each other. In other words, the first pick does not affect the second pick.

The probability of picking a triangle in the first pick is $P(T_1) = \frac{8}{30}$.

The probability of picking a triangle in the second pick is also $P(T_2) = \frac{8}{30}$.

The probability of picking 2 triangular scarves in succession is

$$P(T_1) \cdot P(T_2) = \frac{8}{30} \cdot \frac{8}{30} = \frac{64}{900} = \frac{16}{225}.$$

- The probability of either E or F happening is given by the formula:
$$P(E \text{ or } F) = P(E) + P(F) - P(E \text{ and } F)$$

Example: A card is drawn randomly from a regular deck of cards (52 cards in total, 13 in each of four suits). What is the probability that the card is either a 10 or spades?

Notice that these two sets are overlapping. A card can be a 10 and a spade at the same time. There are a total of four 10s and only one 10 of spades.

$$P(10) = \frac{\# \text{ of } 10\text{s}}{52} = \frac{4}{52} \qquad P(\text{Spade}) = \frac{13}{52} \qquad P(10 \text{ and Spade}) = \frac{1}{52}$$

Add the probabilities of selecting spades and 10s. To avoid double counting, subtract the probability of selecting the 10 of spades.

$$P(10 \text{ or Spade}) = P(10) + P(\text{Spade}) - P(10 \text{ and Spade}) = \frac{4}{52} + \frac{13}{52} - \frac{1}{52} = \frac{16}{52} = \frac{4}{13}$$

MUTUALLY EXCLUSIVE EVENTS

- If two events are mutually exclusive, they cannot both happen. If you pick a dance class at random, it cannot be both a tango and a salsa class at the same time. This means $P(T \text{ and } S) = 0$.
- The probability of either one or the other of two mutually exclusive events happening is given by the following formula:

$$P(T \text{ or } S) = P(T) + P(S) \qquad \text{because} \qquad P(T \text{ and } S) = 0$$

Example: A total of 5 tango classes, 4 salsa classes, and 7 swing classes are offered at a dance studio. If a class is selected randomly, what is the probability that it will be either a swing class or a tango class?

Tango and swing classes are mutually exclusive. Therefore, picking either tango or swing is:

$$P(T \text{ or } S) = P(T) + P(S) = \frac{5}{16} + \frac{7}{16} = \frac{12}{16} = \frac{3}{4}$$

SAMPLE PROBLEMS

EXAMPLE 1

The faces of a cube are marked with the letters *K, K, L, W, M,* and *M*. If the cube is rolled, what is the probability that a *K* will turn up?

A cube has 6 faces, so the number of all possible outcomes is 6.

We want K to show up and there are two Ks, so the number of desired results is 2.

$$P(K) = \frac{2}{6} = \frac{1}{3}$$

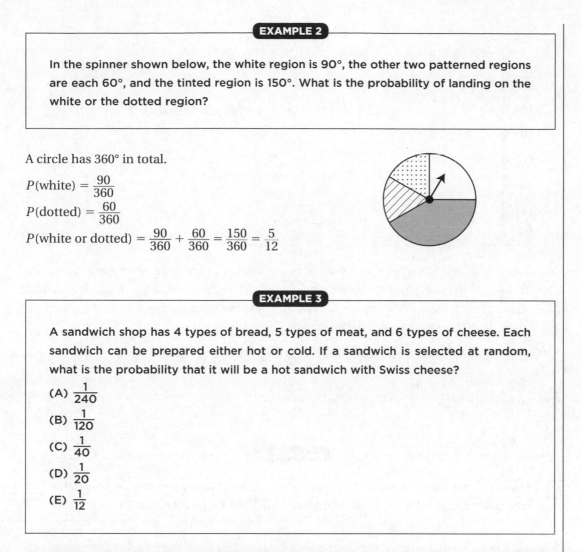

EXAMPLE 2

In the spinner shown below, the white region is 90°, the other two patterned regions are each 60°, and the tinted region is 150°. What is the probability of landing on the white or the dotted region?

A circle has 360° in total.

$P(\text{white}) = \dfrac{90}{360}$

$P(\text{dotted}) = \dfrac{60}{360}$

$P(\text{white or dotted}) = \dfrac{90}{360} + \dfrac{60}{360} = \dfrac{150}{360} = \dfrac{5}{12}$

EXAMPLE 3

A sandwich shop has 4 types of bread, 5 types of meat, and 6 types of cheese. Each sandwich can be prepared either hot or cold. If a sandwich is selected at random, what is the probability that it will be a hot sandwich with Swiss cheese?

(A) $\dfrac{1}{240}$

(B) $\dfrac{1}{120}$

(C) $\dfrac{1}{40}$

(D) $\dfrac{1}{20}$

(E) $\dfrac{1}{12}$

The total number of possible sandwiches can be found by multiplying the number of choices for each selection: $4 \cdot 5 \cdot 6 \cdot 2 = 240$ different sandwiches. So there are 240 possible outcomes. Note that the 2 in the equation stands for the hot versus cold option.

When the desired outcome is a hot sandwich with Swiss cheese, 5 types of meat and 4 types of bread can be chosen. So there are $4 \cdot 5 = 20$ sandwiches that are heated with Swiss cheese (desired outcome).

$P(\text{Swiss and Hot}) = \dfrac{\text{Desired outcomes}}{\text{All possible outcomes}} = \dfrac{20}{240} = \dfrac{1}{12}$

Alternatively, since we are interested in only our choice of cheese and heat, we can ignore the bread and meat choices. We can only pick the cheese $\left(\dfrac{1}{6}\right)$ and the heat $\left(\dfrac{1}{2}\right)$. The probability of getting a hot sandwich with Swiss cheese becomes:

$P(\text{Swiss and Hot}) = \dfrac{1}{6} \cdot \dfrac{1}{2} = \dfrac{1}{12}$

The answer is (E).

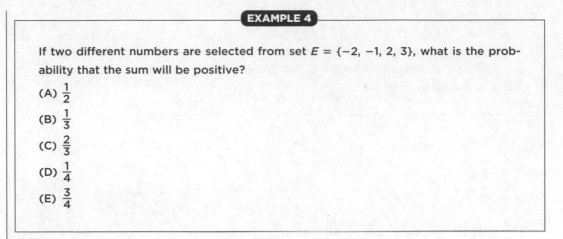

EXAMPLE 4

If two different numbers are selected from set $E = \{-2, -1, 2, 3\}$, what is the probability that the sum will be positive?

(A) $\frac{1}{2}$

(B) $\frac{1}{3}$

(C) $\frac{2}{3}$

(D) $\frac{1}{4}$

(E) $\frac{3}{4}$

There are six different ways to pick a set of two numbers. Since order does not matter, the quickest way is to list all your choices: $(-2, -1), (-2, 2), (-2, 3), (-1, 2), (-1, 3), (2, 3)$. Write your answers systematically so as not to omit any possible pairs. Remember that $(-2, -1)$ is equivalent to $(-1, -2)$, so do not double count.

Out of the 6 choices, 4 have positive sums: $(-2, 3), (-1, 2), (-1, 3)$, and $(2, 3)$.

$P(\text{Positive sum}) = \frac{4}{6} = \frac{2}{3}$

The answer is (C).

EXAMPLE 5

A 6-sided die is rolled twice. The first roll is recorded as m, and the second roll is recorded as k. What is the probability that $\frac{m + k}{m}$ will be greater than 2?

(A) $\frac{5}{12}$

(B) $\frac{1}{2}$

(C) $\frac{1}{3}$

(D) $\frac{21}{36}$

(E) $\frac{17}{36}$

First, simplify the given expression to get more insight on the question.

$\frac{m + k}{m} = \frac{m}{m} + \frac{k}{m} = 1 + \frac{k}{m}$

We want $1 + \frac{k}{m} > 2$,

So subtract 1 from both sides.

$1 + \frac{k}{m} > 2 \quad \rightarrow \quad \frac{k}{m} > 1$

Multiply both sides by m since m is a positive number.

$k > m$

Once we subtract 1 from each side of the inequality, we see that the ratio of k to m needs to be greater than 1. Since both k and m are positive integers, we can conclude that the question is asking for cases where $k > m$ (where the second roll is greater than the first roll).

List all the potential cases. For example, if the first roll is 1, there are 5 cases where the second roll will yield a larger number (2, 3, 4, 5, 6).

First Roll (m)	Second Roll (k)
1	5 possible
2	4 possible
3	3 possible
4	2 possible
5	1 possible
6	m

Desired outcomes $= 5 + 4 + 3 + 2 + 1 = 15$

Total possible outcomes (die rolled twice) $= 6 \cdot 6 = 36$

$P(k > m) = \dfrac{15}{36} = \dfrac{5}{12}$

The answer is (A).

SECTION 2.5—PRACTICE PROBLEMS

1. Set $A = \{2, 4, 8, 16\}$. Set B is formed by multiplying each number in set A by 2. A number is selected randomly from each set to form the fraction $\dfrac{a}{b}$ where a is from set A and b is from set B. What is the probability that the fraction will equal 1?

    (A) $\dfrac{1}{16}$

    (B) $\dfrac{3}{16}$

    (C) $\dfrac{1}{4}$

    (D) $\dfrac{3}{4}$

    (E) 1

2. In a regular deck of cards (52 cards in total, 13 in each of 4 suits, 3 face cards in each suit), if one card is chosen at random, what is the probability that it is a face card or hearts?

    (A) $\dfrac{27}{52}$

    (B) $\dfrac{25}{52}$

    (C) $\dfrac{11}{26}$

    (D) $\dfrac{5}{13}$

    (E) $\dfrac{4}{13}$

3. A keyboard contains only 26 keys (one for each letter in the English alphabet). If 3 keys are randomly pressed, what is the probability of typing MBA?

(A) $\frac{1}{3}$

(B) $\frac{3}{26}$

(C) $\left(\frac{3}{26}\right)^3$

(D) $\frac{1}{26}$

(E) $\left(\frac{1}{26}\right)^3$

4. There are 20 goats, 25 sheep, 3 dogs, and 12 chickens in a barn. If one animal is picked randomly, what is the probability that it will have only 2 legs?

(A) $\frac{1}{20}$

(B) $\frac{1}{5}$

(C) $\frac{1}{3}$

(D) $\frac{5}{12}$

(E) $\frac{4}{5}$

5. A law firm has 15 lawyers that will be randomly assigned to 15 offices, 2 of which are corner offices. If everyone has an equal chance of getting a corner office, what is the probability that both Jasmine and Margaretta will get corner offices?

(A) $\frac{1}{15}$

(B) $\frac{2}{15}$

(C) $\frac{2}{225}$

(D) $\frac{1}{105}$

(E) $\frac{1}{225}$

6. If a number is picked from all positive integers less than 100, what is the probability that it will have 5 in only one of its digits?

(A) $\frac{1}{11}$

(B) $\frac{10}{99}$

(C) $\frac{18}{99}$

(D) $\frac{18}{100}$

(E) $\frac{19}{100}$

7.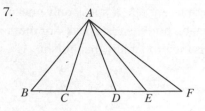

How many triangles are in the figure above?

(A) 6
(B) 9
(C) 10
(D) 11
(E) 12

8. A bag contains blue, yellow, and red balls. The probability of picking a yellow ball is $\frac{1}{6}$. Which of the following cannot be the total number of balls?

(A) 12
(B) 18
(C) 26
(D) 42
(E) 420

9. There are 20 girls and 16 boys in a class. 6 of the girls and 8 of the boys play soccer. If one student is picked randomly, what is the probability that it will be either a soccer player or a girl?

(A) $\frac{7}{18}$
(B) $\frac{7}{9}$
(C) $\frac{3}{18}$
(D) $\frac{5}{9}$
(E) $\frac{4}{9}$

10. Katya has 7 pairs of shoes. In how many different ways can she pick a pair that does not match?

(A) 49
(B) 42
(C) 41
(D) 36
(E) 30

11. Among the attendees of an international marketing conference, 55% speak only one language, 25% speak 2 languages, 15% speak 3 languages, and the rest speak 4 or more languages. If one person is selected randomly, what is the probability that he/she speaks 2 or more languages?

(A) $\frac{1}{4}$

(B) $\frac{2}{5}$

(C) $\frac{9}{20}$

(D) $\frac{1}{18}$

(E) $\frac{7}{18}$

12. 8 people got together for a meeting, and each person shook hands with everybody else. How many handshakes were there?

(A) 8

(B) 16

(C) 28

(D) 56

(E) 64

13. Each letter of the word MISSISSIPPI is written on a piece of paper and placed into a hat. If one piece of paper is selected randomly, what is the probability that it will have an S on it?

(A) $\frac{1}{11}$

(B) $\frac{2}{11}$

(C) $\frac{1}{4}$

(D) $\frac{3}{4}$

(E) $\frac{4}{11}$

14. If two six-sided dice are rolled, what is the probability that the product of the two numbers is odd?

(A) $\frac{1}{4}$

(B) $\frac{1}{9}$

(C) $\frac{1}{16}$

(D) $\frac{1}{18}$

(E) $\frac{1}{36}$

15. Among 50 families surveyed, each owned a car, a house, or both. 28 of the families owned a house, and 42 of the families owned a car. If 1 family is selected randomly, what is the probability that it will own both a house and a car?

(A) $\frac{8}{50}$

(B) $\frac{14}{50}$

(C) $\frac{20}{50}$

(D) $\frac{22}{50}$

(E) $\frac{28}{50}$

16. There are 200 balls that are numbered from 1 to 200. All balls are placed into a bag, and 1 is selected randomly. What is the probability that the number on the ball will be divisible by 6?

(A) $\frac{6}{200}$

(B) $\frac{16}{200}$

(C) $\frac{17}{200}$

(D) $\frac{33}{200}$

(E) $\frac{66}{200}$

17.

A square dartboard is divided into 16 squares as shown. If a dart is thrown randomly and lands on the board, what is the probability that it will land on a dark square?

(A) $\frac{1}{16}$

(B) $\frac{1}{6}$

(C) $\frac{3}{8}$

(D) $\frac{5}{8}$

(E) $\frac{15}{16}$

18. In a race of 8 swimmers, how many different ways could the gold, silver, and bronze medals be distributed?

(A) 216

(B) 336

(C) 340

(D) 343

(E) 512

19. There are red, white, and blue balls in a bag. The probability of picking a red ball is twice as much as picking a white ball and 6 times as much as picking a blue ball. What is the probability of picking a blue ball?

(A) $\frac{1}{10}$

(B) $\frac{1}{5}$

(C) $\frac{2}{9}$

(D) $\frac{1}{3}$

(E) $\frac{1}{2}$

20. A group of six friends wants to have their picture taken with the bride and groom at a wedding ceremony. If the bride and groom stand next to each other and everyone forms a single line, how many different arrangements are possible?

(A) 8!

(B) $7! \cdot 2!$

(C) 7!

(D) $6! \cdot 2!$

(E) 6!

SECTION 2.5—SOLUTIONS

1	2	3	4	5	6	7	8	9	10	11	12	13	14	15	16	17	18	19	20
B	C	E	B	D	C	C	C	B	B	C	C	E	A	C	D	C	B	A	B

1. **B** $A = \{2, 4, 8, 16\}$ and $B = \{4, 8, 16, 32\}$

$4 \cdot 4 = 16$ Fractions can be formed

Three of the fractions equal 1: $\frac{4}{4} = \frac{8}{8} = \frac{16}{16} = 1$.

Probability $= \frac{3}{16}$

2. **C** $P(F) = \frac{\text{\# of face cards}}{52} = \frac{3 \cdot 4}{52} = \frac{12}{52}$

$P(H) = \frac{13}{52}$

These two sets are overlapping. There are 3 cards which are both hearts and face cards. To avoid double counting, subtract the probability of selecting one of these cards.

$P(H \text{ or } F) = P(H) + P(F) - P(H \text{ and } F)$

$\frac{12}{52} + \frac{13}{52} - \frac{3}{52} = \frac{22}{52} = \frac{11}{26}$

3. **E** The number of 3-letter words that can be created from 26 letters is $26 \cdot 26 \cdot 26 = 26^3$.

Only 1 of these words is MBA. $P(MBA) = \left(\frac{1}{26^3}\right)$

Alternatively, build the word one letter at a time.

The probability of pressing M first is $\frac{1}{26}$, of pressing B second is $\frac{1}{26}$, and of pressing A third is $\frac{1}{26}$.

Pressing M, B, and A in order is $\frac{1}{26} \cdot \frac{1}{26} \cdot \frac{1}{26} = \left(\frac{1}{26}\right)^3$.

4. **B** Only the 12 chickens out of $20 + 25 + 3 + 12 = 60$ animals have 2 legs.

$P(2 \text{ legged}) = \frac{12}{60} = \frac{1}{5}$

5. **D** Break this question into two easier parts. Let's assume Jasmine goes first and randomly picks an office. The probability that she will get a corner office is $\frac{2}{15}$.

Once Jasmine picks and gets a corner office, the probability that Margaretta will get a corner office is $\frac{1}{14}$ (1 corner office is left out of 14 offices). Therefore,

$P(J \text{ and } M) = \frac{2}{15} \cdot \frac{1}{14} = \frac{2}{210} = \frac{1}{105}$.

Alternatively, the first corner office can be assigned in 15 different ways. Once it is assigned, the second corner office can be assigned in 14 different ways. Overall, the two corner offices can be assigned in $14 \cdot 15$ different ways.

Our desired outcome is for Jasmine and Margaretta to have corner offices. That can be accomplished 2 different ways:

Corner 1 to M   Corner 1 to J

Corner 2 to J   or   Corner 2 to M

$$P(J \text{ and } F) = \frac{2}{14 \cdot 15} = \frac{1}{105}$$

6. **C** There are 99 positive integers that are less than 100. There are 99 possible outcomes. 18 of these numbers have 5 as only one of its digits (exclude 55).

$$
\begin{array}{c}
5 \\
15 \\
25 \\
35 \\
45 \\
50, 51, 52, 53, 54, \text{—}, 56, 57, 58, 59 \\
65 \\
75 \\
85 \\
95
\end{array}
$$

$$P(5) = \frac{18}{99}$$

7. **C** Count the triangles systematically starting with $\triangle ABC$ to the left.

$\triangle ABC$ $\triangle ABD$ $\triangle ABE$ $\triangle ABF$

$\triangle ACD$ $\triangle ACE$ $\triangle ACF$

$\triangle ADE$ $\triangle ADF$

$\triangle AEF$

There are 10 triangles in total.

8. **C** $P(Y) = \frac{1}{6} = \dfrac{\text{Number of yellow balls}}{\text{Total number of balls}}$

Since the number of balls must be an integer, it has to be a multiple of 6. The only answer choice that is not a multiple of 6 is 26.

9. **B** $P(G) = \dfrac{20}{36}$ Probability of picking a girl is 20 out of 36.

$P(S) = \dfrac{6 + 8}{36} = \dfrac{14}{36}$ Probability of picking a soccer player is 14 out of 36.

$P(S \text{ and } G) = \dfrac{6}{36}$ Probability of picking a female soccer player is 6 out of 36.

$P(S \text{ or } G) = \dfrac{20}{36} + \dfrac{14}{36} - \dfrac{6}{36} = \dfrac{28}{36} = \dfrac{7}{9}$

10. **B** Break down the problem to picking a shoe for the left foot and then for the right foot.

    Katya can pick 7 different left shoes since there are no restrictions yet. After picking the left shoe, she can only pick 6 different shoes for the right foot since she is trying not to match. There are $7 \cdot 6 = 42$ different ways.

11. **C** Since the total number of people is not given, assume there are 100 people to simplify your calculations.

    1 language: 55
    2 languages: 25
    3 languages: 15
    4 or more languages: $100 - (55 + 25 + 15) = 5$

    The number of people who speak 2 or more languages is $25 + 15 + 5 = 45$.

    $P(\text{2 or more}) = \dfrac{45}{100} = \dfrac{9}{20}$

    Alternatively, 2 or more languages means not only 1 language. To find the probability that the person chosen speaks 2 or more languages, subtract the probability of speaking only 1 language from the entire group, $1 - \dfrac{55}{100} = \dfrac{45}{100} = \dfrac{9}{20}$.

12. **C** The first person shook hands with 7 people. The second person shook hands with 6 people because he or she already shook hands with the first person. Make a list of everybody's unique handshakes.

    1st person shook 7 hands
    2nd person shook 6 hands
    3rd person shook 5 hands
    4th person shook 4 hands
    5th person shook 3 hands
    6th person shook 2 hands
    + 7th person shook 1 hand
    _____

    28 handshakes in total

13. **E** There are 11 letters in total, 4 of which are S.

    $P(\text{S}) = \dfrac{4}{11}$

14. **A** For the product to be odd, both numbers need to be odd.

    The probability that the 1st die will show an odd number is $\dfrac{3}{6}$ or $\dfrac{1}{2}$ since there are 3 odd numbers on a die and it has 6 faces. The probability of an odd number for the 2nd die is also $\dfrac{1}{2}$. The probability that both will show an odd number is $P(\text{Odd}) = \dfrac{1}{2} \cdot \dfrac{1}{2} = \dfrac{1}{4}$.

15. **C** Once we add the number of house owners and car owners ($42 + 28 = 70$), we realize that there is an overlap. $70 - 50 = 20$ people own both.

    $P(\text{Both}) = \dfrac{20}{50}$

16. **D** First find out how many numbers between 1 and 200 are divisible by 6.

The smallest number divisible by 6 is 6. The largest number divisible by 6 is 198. It is divisible by 198 because it is divisible by 2 and 3.

$$\frac{198 - 6}{6} = \frac{192}{6} = 32$$

Since we need to count them inclusively, there are $32 + 1 = 33$ numbers.

$$P(6) = \frac{33}{200}$$

17. **C** There are 16 small squares on the board, so the total number of potential outcomes is 16. There are 6 dark squares.

$$P(\text{Dark}) = \frac{6}{16} = \frac{3}{8}$$

18. **B** 8 different swimmers could get the gold. Once one gets the gold, 7 different swimmers can get the silver and then 6 different swimmers can get the bronze.

$$8 \cdot 7 \cdot 6 = 336$$

Alternatively, use the permutation formula;

$$P(8, 3) = \frac{8!}{(8 - 3)!} = \frac{8 \cdot 7 \cdot 6 \cdot 5!}{5!} = 8 \cdot 7 \cdot 6 = 336$$

19. **A** Let the probability of picking blue (B) be x.
The probability of red is 6 times the probability of blue, so $R = 6x$.
The probability of red is 2 times the probability of white, which means the probability of W is half of R, so $W = 3x$.

Since there are only three different colors of balls in the bag, the probabilities must add up to 1.

$$R + W + B = 1$$
$$6x + 3x + x = 1$$
$$10x = 1$$
$$x = \frac{1}{10}$$
$$P(\text{Blue}) = x - \frac{1}{10}$$

20. **B** Since the bride and the groom stand next to each other, consider them as one person initially. If there are 7 people to arrange in a line, there are 7! possible orderings. In addition, the bride and the groom can arrange themselves in 2! different ways (BG or GB). Therefore, the total number of arrangements is 7! · 2!

Algebra

3

→ **3.1 ALGEBRAIC EXPRESSIONS AND FACTORING**

→ **3.2 LINEAR EQUATIONS AND INEQUALITIES**

→ **3.3 EXPONENTS AND RADICALS**

→ **3.4 QUADRATIC EQUATIONS, FUNCTIONS, AND SYMBOLISM**

SECTION 3.1: ALGEBRAIC EXPRESSIONS AND FACTORING

Algebraic Expressions

- In algebra, you work with unknown quantities that are represented by various symbols. These symbols are usually letters of the alphabet. These unknown quantities are called **variables**. They are used as place holders for unknown quantities.
- When letters are placed next to numbers without any operators between them, it means multiplication. Each series (product and/or quotient) of variables and numbers is called a **term**.

"$2x$" means "2 times x" $= 2 \cdot x$

$2xz^2 = 2 \cdot x \cdot z \cdot z$

$3m\sqrt{c} = 3 \cdot m \cdot \sqrt{c}$

$\frac{kr}{m} = \frac{k \cdot r}{m} = k \cdot \frac{r}{m} = \frac{k}{m} \cdot r = k \cdot r \cdot \frac{1}{m}$

- A pattern of variables and numbers is called an **expression**. It is made up of constants, variables, and mathematical operations. An expression may include one or more terms.

$$m - 27 \qquad \text{Expression with two terms: } m \text{ and } 27$$
$$3m + 5y - 4z \qquad \text{Expression with three terms: } 3m, 5y, \text{ and } 4z$$

- An expression with one term is called a **monomial**. The number in front of the variable is called the **coefficient** of the term.

Examples: $2x$, $5xy$, $-17xm^2y^3$, $\left(\frac{1}{5}\right)t$

coefficients

- An expression with more than one term is called a **polynomial**.

    Examples: $n + 3$ $2xz^2 + 5y$ $3(r - k)$ $3x^2 - 5x + 4$

- Since all variables represent a number, all arithmetic rules reviewed in Chapter 2 apply to variables and expressions as well.

ADDING AND SUBTRACTING ALGEBRAIC EXPRESSIONS

> **NOTE**
>
> If there's a negative sign in front of parentheses, it changes the sign of every term that is within the parentheses.
>
> $-(m - 3) = -m + 3$

- Terms that have equivalent variable parts are called **like terms**.

    Examples: $4x$ and $7x$, $27mn^2$ and $7mn^2$,   and $5.5m\sqrt{c}$ and $13m\sqrt{c}$ are like terms.

    $7zt$ and zt are like terms. zt simply means $1zt$.

    $5xy$ and $12yx$ are also like terms since $yx = xy$ (commutative property).

    x and $5x^3$,   and $3xy$ and $3xy^2$ are not like terms.

- Only like terms can be added or subtracted.

    Example: $\underline{3x} + 5y \underline{+ 6x} = 9x + 5y$

    Example: $\boxed{6xy^2} - 5y \boxed{+ 42xy^2} + 7y$ Identify like terms, and mark them.

    $6xy^2 + 42xy^2 - 5y + 7y$ Group like terms together. Remember to move the terms along with their signs.

    $48xy^2 + 2y$ Combine like terms.

    Example: $mn + m^2 - (3mn - 4m^2)$ To remove the parentheses, multiply both $3mn$ and $-4m^2$ by -1.

    $mn + m^2 - 3mn + 4m^2$ Group like terms.

    $mn - 3mn + m^2 + 4m^2$ Add like terms.

    $-2mn + 5m^2$

MULTIPLYING ALGEBRAIC EXPRESSIONS

- When multiplying two terms, multiply the numerical parts together and the variable parts together.

> **NOTE**
>
> $3m \cdot 2m \neq 6m$
>
> $3m \cdot 2m = 6m^2$
>
> $\dfrac{9m}{3m} \neq 3m$
>
> $\dfrac{9m}{3m} = 3$
>
> $3m + 2m = 5m$
>
> $9m - 3m = 6m$

    Examples: $2c \cdot c = 2 \cdot c \cdot 1 \cdot c = 2 \cdot 1 \cdot c \cdot c = 2c^2$

    $3x \cdot 5y = 3 \cdot 5 \cdot x \cdot y = 15xy$

    $5mn \cdot 12m = 5 \cdot 12 \cdot m \cdot m \cdot n = 60m^2n$

- When multiplying two polynomials, use the distributive property.

    Example: $4m \cdot (2 - 5n) = 4m \cdot 2 - 4m \cdot 5n = 8m - 20mn$

    Example: $(m + 2) \cdot (n - 3)$

    $m \cdot (n - 3) + 2 \cdot (n - 3)$ Multiply each term of the first expression by each term of the second expression.

    $mn - 3m + 2n - 6$ There are no like terms, so this is the simplest form of the expression.

DIVIDING ALGEBRAIC EXPRESSIONS

- When dividing two terms, divide the numerical parts with each other and divide/simplify the variable parts with each other.

> **NOTE**
>
> The denominator of a fraction can never be zero since dividing by zero is undefined.

Examples:

$$\frac{2m}{m} = \frac{2 \cdot m}{m} = 2 \qquad \text{Simplify the } m\text{'s since } \frac{m}{m} = 1.$$

$$\frac{16k^4}{8k^2} = \frac{16}{8} \frac{k^4}{k^2} = 2k^2 \qquad \frac{16}{8} = 2 \text{ and } \frac{k^4}{k^2} = \frac{k \cdot k \cdot \cancel{k} \cdot \cancel{k}}{\cancel{k} \cdot \cancel{k}} = k \cdot k = k^2$$

FACTORING AND SIMPLIFYING ALGEBRAIC EXPRESSIONS

- Factors of a term (or a polynomial) are simpler terms that when multiplied together are equal to the original term (or polynomial).

$$2x^2 + x + 5 \qquad\qquad 2 \cdot x \cdot y$$

Terms Factors

Example: Two of the factors of $6m^2$ are $3m$ and $2m$ since $3m \cdot 2m = 6m^2$.
$6m^2 = 2 \cdot 3 \cdot m \cdot m$

- The process of rearranging $ka + kb$ to $k \cdot (a + b)$ is called **factoring** or **taking out the common factor.**

Example: 5 and $(x + 3)$ are factors of $5x + 15$ because $5 \cdot (x + 3) = 5x + 15$.

Example: $ak + at + an = a \cdot (k + t + n)$ — Each term has a factor of a, so a can be factored out of the parentheses.

Example: $6m^2 - 3m + 9m^3 = 3m \cdot (2m - 1 + 3m^2)$ — Each term has a factor of $3m$. Factor out $3m$ from each term. While you are factoring out $3m$, you are dividing each term by $3m$. Therefore, the remaining terms are $\frac{6m^2}{3m} = 2$, $\frac{-3m}{3m} = -1$, and $\frac{9m^3}{3m} = 3m^2$.

Example: $8uv^2 + 4vu^2 - 16uv = 4uv \cdot (2v + u - 4)$ — Each term has a 4, u, and v in it, so $4uv$ can be factored out. The remaining terms are $\frac{8uv^2}{4uv} = 2v$, $\frac{4vu^2}{4uv} = u$, and $\frac{-16uv}{4uv} = -4$.

Example: $\frac{2x - 5}{2x - 5} = 1$ — If the same expression appears in the numerator and the denominator of a fraction, the quotient equals 1 provided that the expression itself is not equal to zero.

Example:

$$\frac{5a^2 - 25a}{a - 5} = \frac{5a(a - 5)}{a - 5}$$

First factor out the common term of $5a$ in the numerator.

$$\frac{5a(a - 5)}{a - 5} = 5a$$

Simplify the term $(a - 5)$ since $\frac{(a - 5)}{(a - 5)} = 1$.

Example:

$$\frac{3x - 6}{2 - x}$$

Factor out a 3 in the numerator.

$$\frac{3(x - 2)}{2 - x}$$

$2 - x$ is equivalent to $-x + 2$ or $-(x - 2)$.

$$\frac{3(x - 2)}{-(x - 2)} = -3$$

Divide out $(x - 2)$, leaving -1 in the denominator.

EVALUATING ALGEBRAIC EXPRESSIONS (SUBSTITUTION)

- An algebraic expression can be evaluated by substituting in known values of all the variables when they are given.

Example: Evaluate $4ab + 3a^2 - 6$ when $a = -2$ and $b = 3$.

Replace all a's with -2 and all b's with 3.

$$4(-2)(3) + 3(-2)^2 - 6$$

Use parentheses for all variables initially so it is clear which variable is replaced by which number. Doing this also helps with determining the order of operations.

$$-24 + 3 \cdot 4 - 6$$

Do the multiplication in the first term and the exponents in the second term. Remember that $(-2)^2 = (-2) \cdot (-2) = 4$.

$$-24 + 12 - 6 = -18$$

SAMPLE PROBLEMS

EXAMPLE 1

Multiply $(a + 2)$ by $(a - 7)$.

$(a + 2) \cdot (a - 7)$

Multiply each term of the first expression by each term of the second expression.

$a \cdot a - 7a + 2a - 14$

Multiply, add, and subtract like terms.

$a^2 - 5a - 14$

EXAMPLE 2

Simplify $x(x - 5) - 16(x - 5)$.

$x(x - 5) - 16(x - 5)$ Notice that $(x - 5)$ is a common factor, so it can be factored out.

$$x(x - 5) - 16(x - 5)$$

$$(x - 5)(x - 16)$$

EXAMPLE 3

Factor $6m^2n - 12mn + 18mn^2$.

$6m^2n - 12mn + 18mn^2$ Notice that $6mn$ is a common factor. Factor out $6mn$.

$6mn(m - 2 + 3n)$ The remaining terms are

$$\frac{6m^2n}{6mn} = m, \quad \frac{-12mn}{6mn} = -2, \quad \text{and} \quad \frac{18mn^2}{6mn} = 3n.$$

SECTION 3.1—PRACTICE PROBLEMS

1. What is the simplest form of $2(x + 3) - 3(x - 2)$?

   (A) 10
   (B) 12
   (C) $8x$
   (D) $12 - x$
   (E) $x + 12$

2. $\dfrac{2mk - m}{1 - 2k} = ?$

   (A) $-k$
   (B) $-m$
   (C) 1
   (D) m
   (E) k

3. If $x = -2$, what is the value of $-2x^2 - 3x^3$?

   (A) -24
   (B) -16
   (C) -8
   (D) 16
   (E) 32

4. Which of the following is equivalent to $(k + t)(3k + 5) - 2k - 2t$?

   (A) $3(k + t)(k + 1)$
   (B) $(k + t)(k + 3t)$
   (C) $(k - t)(3k + 5)$
   (D) $3(k - t)(k + t)$
   (E) $(3k + 5)(k + 2t)$

5. The difference of $(x + 1)^2$ and $(x - 1)^2$ is how many times greater than x?

   (A) -2
   (B) 2
   (C) 4
   (D) $3x$
   (E) $2x + 4$

6. What is the value of the expression $-2m^2n - 3mn^2$ when $m = -1$ and $n = -2$?

   (A) -16
   (B) -8
   (C) 4
   (D) 8
   (E) 16

7. The expression $(x - 3)(2x + 5) - x(2x - 1)$ is equivalent to which of the following?

   (A) $-15 - 2x$
   (B) $-2x$
   (C) -15
   (D) $x^2 - 2x - 15$
   (E) $-15 + 2x$

8. $\dfrac{xyz + x^2yz + xy^2z + xyz^2}{xyz} = ?$

   (A) xyz
   (B) $4xyz$
   (C) $x + y + z$
   (D) $1 + x + y + z$
   (E) $1 + x^2 + y^2 + z^2$

9. What is the value of $\dfrac{can - cn}{nc}$ if $a = 5$?

   (A) 4
   (B) 3
   (C) 2
   (D) 1
   (E) It cannot be determined from the information given.

10. $(7x - 2)(2x + 1) - (13x^2 + 3x - 2) = ?$

   (A) x
   (B) $x(x - 6)$
   (C) $27x^2 - 4$
   (D) x^2
   (E) $x^2 - 4$

11. $(n - 2)(x + 3)$ is how much more than $-(4 - n)(x + 3)$?

    (A) $2nx + 6n + 2x+6$

    (B) $2nx + 2x + 6$

    (C) $2x + 6$

    (D) 6

    (E) 2

12. If $2x + 3y = 12$, what is the value of $\frac{2x}{3} + y + 4x^2 + 12xy + 9y^2 - 4x - 6y$?

    (A) 4

    (B) 16

    (C) 44

    (D) 124

    (E) It cannot be determined from the information given.

SECTION 3.1—SOLUTIONS

1	2	3	4	5	6	7	8	9	10	11	12
D	B	D	A	C	E	C	D	A	D	C	D

1. **D** $2(x + 3) - 3(x - 2)$ Distribute 2 and -3.

   $2x + 6 - 3x + 6$ Add and subtract like terms.

   $-x + 12 = 12 - x$

2. **B** $\dfrac{2mk - m}{1 - 2k}$ Factor out an m in the numerator.

   $\dfrac{m(2k - 1)}{1 - 2k}$ Rewrite $1 - 2k$ as $-(2k - 1)$,

   $\dfrac{m(2k - 1)}{-(2k - 1)} = -m$

3. **D** Substitute -2 for each x. It is a good idea to use parentheses in the first step.

   $-2(-2)^2 - 3(-2)^3$ $(-2)^2 = 4$ $(-2)^3 = -8$

   $-2(4) - 3(-8)$ Multiply.

   $8 + 24 = 16$

4. **A** $(k + t)(3k + 5) - 2k - 2t$ Factor out -2 from $-2k - 2t$.

   $(k + t)(3k + 5) - 2(k + t)$ $(k + t)$ becomes a common term. Factor it out.

   $(k + t)(3k + 5 - 2)$

   $(k + t)(3k + 3)$ Factor out a 3.

   $3(k + t)(k + 1)$

5. **C** First find the difference of $(x + 1)^2$ and $(x - 1)^2$.

   $(x + 1)^2 - (x - 1)^2$ Square and subtract or use the difference of two squares formula from Section 3.4.

   $x^2 + 2x + 1 - (x^2 - 2x + 1)$

   $x^2 + 2x + 1 - x^2 + 2x - 1$

   $= 4x$

   $\dfrac{4x}{x} = 4$ Divide $4x$ by x to find how many times greater the difference is than x.

6. **E** Plug in -1 for m and -2 for n.

$$-2(-1)^2(-2) - 3(-1)(-2)^2$$
$$-2 \cdot 1 \cdot (-2) - 3(-1) \cdot (4)$$
$$4 + 12 = 16$$

7. **C** $(x - 3)(2x + 5) - x(2x - 1)$ — Multiply since no common terms can be factored out.

$2x^2 + 5x - 6x - 15 - 2x^2 + x$ — Add and subtract like terms.
$2x^2 - 2x^2 + 5x - 6x + x - 15 = -15$

8. **D** Notice that xyz is a common term in the numerator.

$$\frac{xyz + x^2yz + xy^2z + xyz^2}{xyz}$$ Factor out xyz first.

$$\frac{xyz(1 + x + y + z)}{xyz}$$ Simplify xyz.

$$= 1 + x + y + z$$

9. **A** The numerator is $c \cdot a \cdot n - c \cdot n$. So cn is a common factor that can be factored out.

$$\frac{cn(a - 1)}{nc} = a - 1$$

Plug in 5 for a.
$a - 1 = (5) - 1 = 4$

10. **D** $(7x - 2)(2x + 1) - (13x^2 + 3x - 2)$ — Multiply the first term.

$14x^2 + 7x - 4x - 2 - (13x^2 + 3x - 2)$ — Remove the parentheses by multiplying each term by -1.

$14x^2 + 7x - 4x - 2 - 13x^2 - 3x + 2$ — Group the like terms, and then add and subtract.

$14x^2 - 13x^2 + 3x - 3x - 2 + 2 = x^2$

11. **C** Find the difference between the two terms by subtracting the second from the first.

$(n - 2)(x + 3) - [-(4 - n)(x + 3)]$ — Multiply the terms. Two negatives will turn into an addition.

$nx + 3n - 2x - 6 + 4x + 12 - nx - 3n$ — Group the like terms, and then add and subtract.

$nx - nx + 3n - 3n - 2x + 4x - 6 + 12$
$2x + 6$

Alternatively,
$(n - 2)(x + 3) + (4 - n)(x + 3)$ — Factor out the $(x + 3)$.
$(x + 3)(n - 2 + 4 - n)$
$(x + 3)(2) = 2x + 6$

12. **D** Notice that the certain terms in the given equation can be written in terms of $2x + 3y$, which equals 12.

$$\underbrace{\frac{2x}{3} + y}_{A} + \underbrace{4x^2 + 12xy + 9y^2}_{B} \underbrace{- 4x - 6y}_{C}$$

A: $\quad \dfrac{2x}{3} + y = \dfrac{2x + 3y}{3} = \dfrac{12}{3} = 4$

B: $\quad 4x^2 + 12xy + 9y^2 = (2x + 3y)^2 = 12^2 = 144$

C: $\quad -4x - 6y = -2(2x + 3y) = -2 \cdot 12 = -24$

$\dfrac{2x}{3} + y + 4x^2 + 12xy + 9y^2 - 4x - 6y = 4 + 144 - 24 = 124$

SECTION 3.2: LINEAR EQUATIONS AND INEQUALITIES
Linear Equations

SOLVING A LINEAR EQUATION WITH ONE UNKNOWN

- An equation states that two algebraic expressions are equal. For example, "sixteen more than 2 times a number is equal to 5 less than half the same number" is an equation in plain English that translates into $2n + 16 = \dfrac{n}{2} - 5$.

- The **solution** of an equation is the value of all stated variables that make the statement true. In other words, the solution is all numbers that satisfy the equation.

    Example: In the equation $3m + 2 = 8$, replacing m with 2 gives you $3(2) + 2 = 8$, which is true. Therefore, 2 is the solution of the equation $3m + 2 = 8$.

- To solve an equation with one unknown, isolate the variable on one side of the equation. Isolating a variable means it remains alone on one side of the equation. When the variable is alone, it has a coefficient of 1. The side the variable is on does not matter.

    Examples: $n = -3$

    $\dfrac{2}{3} = x \left(\text{which is the same as } x = \dfrac{2}{3} \right)$

- To isolate a variable, use the following properties.

    Addition/Subtraction: You can add (or subtract) the same quantity to (or from) each side of the equation without affecting the equation.

    Examples:

$x = 6$	$n - 4 = 5$	$n + 2 = 5$
$x + 3 = 6 + 3$	$n - 4 + 4 = 5 + 4$	$n + 2 - 2 = 5 - 2$
$x + 3 = 9$	$n = 9$	$n = 3$

    Multiplication/Division: You can multiply (or divide) each side of the equation by the same nonzero quantity without affecting the equation.

    Examples:

$x = 6$	$\dfrac{n}{4} = 5$	$7n = 21$
$3 \cdot x = 3 \cdot 6$	$4 \cdot \dfrac{n}{4} = 5 \cdot 4$	$\dfrac{7n}{7} = \dfrac{21}{7}$
$3x = 18$	$n = 20$	$n = 3$

THREE-STEP PROCESS TO SOLVE LINEAR EQUATIONS WITH ONE UNKNOWN

1. EVALUATE each side of the equation as much as possible by multiplying exponents, simplifying parentheses, and combining like terms.
2. MOVE all variables to one side and all numbers to the other side of the equal sign using the properties previously described, and simplify both sides again.
3. ISOLATE the desired variable.

Example: $5 + 4(n - 2) + 3(n + 5) = 2(n - 7) + 1$

1. Evaluate

$5 + 4(n - 2) - 3(n + 5) = 2(n - 7) + 1$	Use the distributive property to remove the parentheses.
$5 + 4n - 8 - 3n - 15 = 2n - 14 + 1$	Group like terms together.
$4n - 3n + 5 - 8 - 15 = 2n - 14 + 1$	Combine like terms.
$n - 18 = 2n - 13$	Once each side is simplified, go to steps 2 and 3.

2. Move

$n - 18 = 2n - 13$	Add 18 to each side of the equation so the numbers are all on the right side.
$n = 2n + 5$	Subtract $2n$ from each side so the variables are all on the left.
$-n = 5$	

3. Isolate

$-n = 5$	Multiply each side by -1.
$n = -5$	

- If you end up with a false statement at the end of your solution, either the equation does not have a solution (does not work for any number) or you have made a mistake.

Example:

$3(x + 4) - 16 = 27 + 3x$	Distribute 3 on the left side.
$3x + 12 - 16 = 27 + 3x$	Add like terms.
$3x - 4 = 27 + 3x$	Subtract $3x$ from both sides.
$-4 = 27$, which is false	This means the original equation has no solution.

- If you end up with a true statement at the end but no variables, the equation works for all real numbers.

Example:

$3(x + 4) - 16 = 3x - 4$	Distribute 3 on the left side.
$3x + 12 - 16 = 3x - 4$	Add like terms.
$3x - 4 = 3x - 4$	Subtract $3x$ from both sides.
$-4 = -4$, which is true	This means the equation is true for any real number.

SOLVING TWO LINEAR EQUATIONS WITH TWO UNKNOWNS

- Two independent equations are needed to solve for two unknowns.

    Example: $2x + 3y = 6$
    $\frac{x}{2} - 5y = 11$

- Two common methods can be used to solve these equations simultaneously, substitution and elimination.

Substitution:

1. Use one of the equations to isolate one of the variables. Pick the variable that is easier to isolate.

2. Substitute the isolated variable into the second equation to transform it into a single linear equation with one unknown.

3. Solve the equation using the three-step process described previously.

Example: $2x + 3y = 11$
$5x - y = 2$

Isolate the y in the second equation. It is relatively easy to isolate since it has a coefficient of -1.

$5x - y = 2$	Subtract $5x$ from both sides.
$-y = 2 - 5x$	Multiply both sides by -1.
$y = -2 + 5x$	Substitute this y into the first equation as follows.
$2x + 3y = 11$	
$2x + 3(-2 + 5x) = 11$	Distribute the 3.
$2x - 6 + 15x = 11$	Add the like terms.
$17x - 6 = 11$	Add 6 to both sides.
$17x = 17$	Divide both sides by 17.
$x = 1$	

After solving for one of the variables, substitute it into one of the original equations to solve for the second variable. Use the first equation for example.

$2x + 3y = 11$	Plug in 1 for x.
$2(1) + 3y = 11$	Subtract 2 from both sides.
$3y = 9$	Divide both sides by 3.
$y = 3$	

This means the pair $x = 1$ and $y = 3$ is the solution of both equations. Check your answer by substituting these values into the original equations.

$2x + 3y = 11$	$2(1) + 3(3) = 11$	$11 = 11 ✓$
$5x - y = 2$	$5(1) - 3 = 2$	$2 = 2 ✓$

Elimination:

- Elimination is based on the fact that two equations can be added together to create a third equation.

$$2x + 4y = 13$$
$$\underline{+\ 3x - 2y = -5}$$
$$5x + 2y = 8$$

- To eliminate one of the variables, the coefficients of the variable you are trying to eliminate should have the same absolute value. If they do not, you can easily multiply one or both of the equations by a constant number to make them the same.

Example:

$$2x + 3y = 6$$
$$\underline{7x - 3y = 12}$$
$$9x + 0 = 18$$

Since the coefficients of y are 3 and -3, adding these equations side by side will eliminate y.

$$9x = 18$$ Divide both sides by 9.
$$\underline{x = 2}$$ Plug $x = 2$ into the first equation.
$$2(2) + 3y = 6$$ Subtract 4 from each side.
$$3y = 6 - 4$$ Divide both sides by 3.
$$y = \frac{2}{3}$$

Example:

$$2x + 3y = 14$$
$$-x + 5y = 6$$

If you multiply the second equation by 2, x can be eliminated using addition.

$$2 \cdot (-x + 5y) = 2 \cdot (6)$$ Multiply every term by 2.
$$-2x + 10y = 12$$ Add this equation to the first equation to eliminate x.

$$2x + 3y = 14$$
$$\underline{-2x + 10y = 12}$$
$$0 + 13y = 26$$

$$13y = 26$$ Divide both sides by 13.
$$y = 2$$ Plug 2 into the first equation for y.
$$2x + 3(2) = 14$$ Subtract 6 from each side.
$$2x = 8$$ Divide both sides by 2.
$$x = 4$$

- If two equations are multiples of each other, there are infinitely many solutions. In other words, an infinite number of (x, y) pairs satisfy the equation.

Example:

$$3x - 5y = 21$$
$$9x - 15y = 63$$

If you multiply the first equation by 3, $3 \cdot (3x - 5y = 21) = 9x - 15y = 63$, you get the second equation. This means these two equations are equivalent. In this case, there are infinitely many (x, y) pairs that satisfy both equations, such as $x = 2$, $y = -3$ or $x = 7$, $y = 0$.

If you do not notice this from the beginning and attempt to solve these types of equations, you get a true statement with no variables.

$$3x - 5y = 21$$ becomes $$9x - 15y = 63$$
$$9x - 15y = 63$$ becomes $$\underline{9x - 15y = 63}$$ Subtract the equations.
                                          $$0 = 0$$ The result is true, which means there are infinitely many solutions.

- If the equations present contradicting results, there is no solution.

Example:

$$3x - 5y = 21$$
$$9x - 15y = 62$$

If you multiply the first equation by 3, $3 \cdot (3x - 5y = 21)$ you get $9x - 15y = 63$. The left-hand side becomes the same as that of the second equation, but it is equal to a different constant. This means these two equations present contradicting information. In these cases, no (x, y) pairs satisfy both equations.

If you do not notice this from the beginning and attempt to solve these types of equations, you get a false statement with no variables.

$$3x - 5y = 21 \quad \text{becomes} \quad 9x - 15y = 63$$
$$\underline{9x - 15y = 62 \quad \text{becomes} \quad 9x - 15y = 62} \quad \text{Subtract both sides.}$$
$$0 = 1 \quad \text{The result is false, which means there is no solution.}$$

Linear Inequalities

- An inequality compares two algebraic expressions. For example, "three times a number is greater than 5 more than twice the same number" is an inequality in plain English that translates into $3n > 2n + 5$.

SIGNS OF INEQUALITY

$\neq$	not equal to
$>$	greater than
$\geq$	greater than or equal to
$<$	less than
$\leq$	less than or equal to

- Solving a linear inequality with one variable is essentially the same as solving an equation with one variable. The only difference is that when you multiply or divide both sides of an inequality by a negative, nonzero number, the inequality sign reverses. Everything else is the same as working with equations.

Example:

$$3 - 2x \geq 5 \quad \text{Subtract 3 from both sides.}$$
$$\underline{-3 \qquad -3}$$
$$-2x \geq 2 \quad \text{Divide both sides by } -2 \text{ and therefore reverse}$$
$$x \leq -1 \quad \text{the inequality sign.}$$

Example:

$$-4 \geq \frac{4x - 4}{-5} \qquad \text{Multiply both sides by } -5 \text{ and reverse the inequality.}$$
$$(-5) \cdot (-4) \leq \left(\frac{4x - 4}{-5}\right) \cdot (-5) \qquad \text{Simplify the } -5\text{s on the right side.}$$
$$20 \leq 4x - 4 \qquad \text{Add 4 to both sides.}$$
$$24 \leq 4x \qquad \text{Divide both sides by 4.}$$
$$6 \leq x \qquad x \text{ can be any real number greater than or equal to 6.}$$

REMEMBER

When you multiply or divide both sides of an inequality by a negative number, the inequality sign reverses.

■ Two inequalities can be added side by side if the inequality signs have the same direction.

$$a < b$$
$$+ \quad c < b$$
$$\overline{a + c < b + d}$$

SAMPLE PROBLEMS

EXAMPLE 1

Find x if $4 + 3(x - 4) = -x - (2x - 16)$

(A) $\frac{3}{4}$

(B) 1

(C) $\frac{4}{3}$

(D) 3

(E) 4

$4 + 3(x - 4) = -x - (2x - 16)$ Evaluate the contents of the parentheses first. Distribute 3 and -1.

$4 + 3x - 12 = -x - 2x + 16$ Add and subtract like terms.

$3x - 8 = -3x + 16$ Add $3x$ to both sides.

$6x - 8 = 16$ Add 8 to both sides.

$6x = 24$ Divide both sides by 6.

$x = 4$

The answer is (E).

EXAMPLE 2

Solve for x and y if $x - 2y = 12$ and $2x - 3y = 25$.

Use substitution since x can be easily isolated in the first equation.

$x - 2y = 12$ Add $2y$ to both sides.

$x = \underline{12 + 2y}$ Substitute the equivalent of x from this equation into the second equation.

$2(x) - 3y = 25$

$2(12 + 2y) - 3y = 25$ Distribute 2.

$24 + 4y - 3y = 25$ Add like terms.

$24 + y = 25$ Subtract 24 from both sides.

$y = 1$

Substitute $y = 1$ into the first equation to find x.

$x - 2(1) = 12$

$x - 2 = 12$

$x = 14$

If 3 lbs of apples and 5 lbs of oranges cost \$3.15, 15 lbs of oranges and 9 lbs of apples cost \$9.45. How much does 1 lb. of apples cost?

(A) \$1.10

(B) \$1.05

(C) \$0.90

(D) \$0.80

(E) It cannot be determined from the information given.

First translate the two equations.

3 lbs of apples and 5 lbs of oranges cost \$3.15   translates as $3A + 5R = 3.15$

15 lbs of oranges and 9 lbs of apples cost \$9.45   translates as $9A + 15R = 9.45$

Use elimination since neither A nor R can easily be isolated. Multiply the first equation by 3 to match the coefficients of R and then subtract.

$3 \cdot (3A + 5R = 3.15) \cdot 3$

$9A + 15R = 9.45$ Bring over the second equation and subtract.

$\underline{9A + 15R = 9.45}$

           $0 = 0$ True   These two equations present the same information.

There are infinitely many solutions. Therefore not enough information is provided to solve the problem.

The answer is (E).

SECTION 3.2—PRACTICE PROBLEMS

1. Find x if $\frac{x}{5} - \frac{2x + 1}{3} + \frac{4 - x}{15} = \frac{8 - 7x}{15}$.

   (A) -15

   (B) -12

   (C) -9

   (D) -5

   (E) -3

2. If $3(a - 1) + 2(2a - 3) - 3(2a - 3) + a$, what does a equal?

   (A) No solution

   (B) -1

   (C) 0

   (D) 1

   (E) All real numbers

3. If $\frac{m}{m - 3} = 6 - \frac{7}{m - 3}$, what does m equal?

   (A) 3

   (B) 4

   (C) 5

   (D) 6

   (E) 7

4. If $-3x - 2 \geq -\frac{1}{2}$, what is the greatest integer value of x?

   (A) -2
   (B) -1
   (C) 0
   (D) 1
   (E) 2

5. Three adults with two children paid $127 in total for concert tickets. If a child's ticket costs less than an adult's ticket and if three children and two adults pay a total of $113, how much are the tickets for one adult with one child?

   (A) $40
   (B) $48
   (C) $50
   (D) $60
   (E) $80

6. If $0 < a < b$ and $k = \dfrac{2a + 7b}{b}$, which of the following must be true?

   (A) $k < 2$
   (B) $k < 7$
   (C) $k < 9$
   (D) $k > 9$
   (E) $k > 11$

7. If $-4 < a < 2$ and $-6 < b < 3$, what is the largest integer value of $a^2 + b^2$?

   (A) 12
   (B) 34
   (C) 45
   (D) 51
   (E) 52

8. What is the value of x if $x - 2x + 3x - 4x + 5x - 6x + \ldots + 11x - 12x = -18$?

   (A) -3
   (B) -1
   (C) 0
   (D) 1
   (E) 3

9. If $\dfrac{1}{9} < \dfrac{2}{x - 6} < \dfrac{1}{3}$, how many integer values can x take?

   (A) 6
   (B) 8
   (C) 10
   (D) 11
   (E) 12

10. If $\dfrac{2}{x-2} = \dfrac{-4}{4-2x}$, which of the following cannot be the value of x?

    (A) 0
    (B) 1
    (C) 2
    (D) 3
    (E) 4

11. $2x + 3y + 1 = 0$ and $-1 < y < 2$. What is the greatest integer value of x?

    (A) -3
    (B) -2
    (C) 0
    (D) 1
    (E) 2

12. If $A - 4x + 3$ and $B = 3x - 4$, which x results in the smallest value that $|A - B|$ can take?

    (A) -7
    (B) -1
    (C) 0
    (D) 1
    (E) 7

SECTION 3.2—SOLUTIONS

1	2	3	4	5	6	7	8	9	10	11	12
C	E	C	B	B	C	D	E	D	C	C	A

1. **C**

$$15\left(\dfrac{x}{5} - \dfrac{2x+1}{3} + \dfrac{4-x}{15}\right) = \left(\dfrac{8-7x}{15}\right)15$$

Multiply both sides of the equation by 15 to simplify the denominators.

$$\left(15 \cdot \dfrac{x}{5} - 15 \cdot \dfrac{2x+1}{3} + 15 \cdot \dfrac{4-x}{15}\right) = \left(\dfrac{8-7x}{15} \cdot 15\right)$$

Multiply each term by 15 and simplify.

$$3x - 5(2x+1) + 4 - x = 8 - 7x$$

Distribute -5.

$$3x - 10x - 5 + 4 - x = 8 - 7x$$

Simplify.

$$-8x - 1 = 8 - 7x$$

Add $8x$.

$$-1 = 8 + x$$

Subtract 8.

$$-9 = x$$

2. **E** $3(a - 1) + 2(2a - 3) = 3(2a - 3) + a$

First distribute 3 and 2 on the left and distribute 3 on the right.

$$3a - 3 + 4a - 6 = 6a - 9 + a$$

Add and subtract like terms.

$$7a - 9 = 7a - 9$$

Subtract $7a$ and add 9.

$$0 = 0$$

Since you ended up with a true statement, there are infinitely many solutions. In this case, the equation is satisfied for all real values of a.

3. **C** Since $\dfrac{m}{m-3}$ and $\dfrac{7}{m-3}$ already have the same denominator, add $\dfrac{7}{m-3}$ to both sides.

$$\dfrac{m}{m-3} + \dfrac{7}{m-3} = 6 \qquad \text{Add the fractions.}$$

$$\dfrac{m+7}{m-3} = \dfrac{6}{1} \qquad \text{Cross multiply.}$$

$$6 \cdot (m-3) = 1(m+7) \qquad \text{Distribute.}$$

$$6m - 18 = m + 7 \qquad \text{Subtract } m \text{ from both sides.}$$

$$5m - 18 = 7$$

$$5m = 25$$

$$m = 5$$

4. **B** $\qquad -3x - 2 \geq -\dfrac{1}{2} \qquad$ Multiply both sides by -2, and remember to reverse the inequality sign.

$$(-3x - 2) \cdot -2 \leq \left(-\dfrac{1}{2}\right) \cdot -2$$

$$6x + 4 \leq 1 \qquad \text{Subtract 4.}$$

$$6x \leq -3 \qquad \text{Divide by 6.}$$

$$x \leq -\dfrac{1}{2} \qquad \text{The greatest integer that satisfies } x \leq -\dfrac{1}{2} \text{ is } -1.$$

5. **B** Translate the statements first.

Three adults with two children paid $127 is $3A + 2C = 127$.
Three children and two adults paid $113 is $2A + 3C = 113$.
The question is $A + C = ?$

When a combination of x and y (such as $x + y$) is asked instead of x and y individually, it is often quicker to get to the result directly by adding or subtracting.

$$\begin{array}{r} 3A + 2C = 127 \\ + \ 2A + 3C = 113 \\ \hline 5A + 5C = 240 \end{array} \qquad \text{Divide both sides by 5.}$$

$$A + C = 48$$

6. **C** Rewrite $\dfrac{2a + 7b}{b}$ as $\dfrac{2a}{b} + \dfrac{7b}{b}$, which simplifies to $\dfrac{2a}{b} + 7$.

Since $0 < a < b$, $\dfrac{a}{b}$ must be less than 1 since b is greater than a: $\dfrac{a}{b} < 1$. Multiply both sides by 2 to get $\dfrac{2a}{b} < 2$.

If $k = \dfrac{2a}{b} + 7$ and $\dfrac{2a}{b} < 2$, k must be less than 9.

7. **D** Since squares of negative numbers are positive, pick an a and a b with the greatest absolute value from each interval. Since $-4 < a$ and $-6 < b$, pick $a = -4$ and $b = -6$ as limiting values.

Your choice for a and b do not need to be integers. They can be any real numbers that are close to -4 and -6 to make $a^2 + b^2$ both as large as possible and an integer.

$(-4)^2 + (-6)^2 = 16 + 36 = 52$. You know that $a \neq -4$ and $b \neq -6$. Pick the next integer that is less than 52, which is 51.

8. **E** Group the x's on the left so that the sum of each group is the same for each calculation.

$$\underbrace{x - 2x}_{-x} + \underbrace{3x - 4x}_{-x} + \underbrace{5x - 6x}_{-x} + \underbrace{7x - 8x}_{-x} + \underbrace{9x - 10x}_{-x} + \underbrace{11x - 12x}_{-x} = -18$$
$$= -18$$
$$-6x = -18$$
$$x = 3$$

9. **D** Expand the first and the third fraction by 2 to make the numerators 2 for easy comparison:

$$\frac{2}{18} < \frac{2}{x - 6} < \frac{2}{6}$$

$(x - 6)$ must be a number between 6 and 18. There are 11 numbers between 6 and 18 $(18 - 6 - 1 = 11)$.

If $(x - 6)$ can take 11 different integer values, then x can also take 11 different integer values.

10. **C** $\dfrac{2}{x - 2} = \dfrac{-4}{4 - 2x}$ Cross multiply.

$2(4 - 2x) = -4(x - 2)$ Distribute.

$8 - 4x = -4x + 8$ Add $4x$ and subtract 8.

$0 = 0$ True statement.

The solution is all real numbers except 2 because 2 makes the denominators equal zero. Denominators cannot be zero since any number divided by zero is undefined.

Alternatively, check the denominators of each fraction before you start. Remember that $x - 2$ and $4 - 2x$ cannot equal zero because dividing by zero is undefined.

$$x - 2 \neq 0$$
$$x \neq 2$$

and

$$4 - 2x \neq 0$$
$$-2x \neq -4$$
$$x \neq 2$$

11. **C** The easiest way to approach this problem is to try answer choices for x (starting from choice (E) since you are looking for the greatest value) and see which one results in a y that is between -1 and 2.

Try $x = 2$ $2(2) + 3y + 1 = 0$ $4 + 3y + 1 = 0$ $y = -\dfrac{5}{3}$ Not in the range

Try $x = 1$ $2(1) + 3y + 1 = 0$ $2 + 3y + 1 = 0$ $y = -\dfrac{3}{3} = -1$ Not in the range

Try $x = 0$ $2(0) + 3y + 1 = 0$ $0 + 3y + 1 = 0$ $y = -\dfrac{1}{3}$ In the range ✓

So $x = 0$.

12. **A** $|A - B| = |4x + 3 - (3x - 4)| = |4x + 3 - 3x + 4| = |x + 7|$

$|x + 7|$ cannot be negative. The least value it can take is 0.

Make $(x + 7)$ equal to zero to solve for x.

$$x + 7 = 0$$
$$x = -7$$

SECTION 3.3: EXPONENTS AND RADICALS
Exponents

- When a number k is multiplied by itself n times, it is represented as k^n. k is called the base, and n is called the **exponent**.

  $2 \cdot 2 \cdot 2 \cdot 2 \cdot 2 = 2^5$

  $k \cdot k \cdot k \cdot k = k^4$

- $x^1 = x$ Any number to the power of 1 equals itself.

  $42^1 = 42$ $7^1 = 7$

- $x^0 = 1$ Any nonzero number to the power of 0 equals 1.

  $13^0 = 63^0 = 2^0 = 1$

  $0^0 =$ undefined

- If $x^m = x^n$ then $m = n$ except when $x = -1, 0,$ or 1.

  If $12^{m+1} = 12^8$,   then $m + 1 = 8$ and $m = 7$.

- All powers of positive numbers are positive.

  $7^3 = 343$

  $7^2 = 49$

- Even powers of negative numbers are positive, and odd powers of negative numbers are negative.

  $(-7)^3 = -343$

  $(-7)^2 = 49$

MEMORIZE

$2^1 = 2$	$3^3 = 27$	$3^2 = 9$
$2^2 = 4$	$4^3 = 64$	$4^2 = 16$
$2^3 = 8$	$5^3 = 125$	$5^2 = 25$
$2^4 = 16$		$6^2 = 36$
$2^5 = 32$		$7^2 = 49$
$2^6 = 64$		$8^2 = 64$
$2^7 = 128$		$9^2 = 81$
$2^8 = 256$		$10^2 = 100$
		$11^2 = 121$
		$12^2 = 144$
		$13^2 = 169$
		$14^2 = 196$
		$15^2 = 225$

MULTIPLICATION AND DIVISION

- To multiply when the bases are the same, add the exponents.

  $x^a \cdot x^b = x^{a+b}$

- To divide when the bases are the same, subtract the exponent of the denominator from the exponent of the numerator.

  $\dfrac{x^a}{x^b} = x^{a-b}$

Examples:
$$4^2 \cdot 4^{12} = 4^{14}$$
$$5^5 \div 5^3 = 5^2$$
$$6^7 \cdot 6 = 6^7 \cdot 6^1 = 6^8$$
$$\frac{5^5}{5} = \frac{5^5}{5^1} = 5^{5-1} = 5^4$$

- To multiply when the exponents are the same, multiply the bases and keep the exponent same.

$$x^a \cdot y^a = (x \cdot y)^a$$

- To divide when the exponents are the same, divide the bases and keep the exponent same.

$$\frac{x^a}{y^a} = \left(\frac{x}{y}\right)^a$$

Examples:
$$2^3 \cdot x^3 = (2x)^3$$
$$(12)^2 = (2 \cdot 2 \cdot 3)^2 = 2^2 \cdot 2^2 \cdot 3^2 \text{ or } 4^2 \cdot 3^2$$
$$\frac{3^3}{7^3} = \left(\frac{3}{7}\right)^3$$

NEGATIVE EXPONENTS

- Negative exponents take the reciprocal of the base and then become positive.

$$x^{-n} = \left(\frac{1}{x}\right)^n = \frac{1}{x^n}$$

$$\left(\frac{y}{x}\right)^{-n} = \left(\frac{x}{y}\right)^n$$

Examples:
$$5^{-2} = \left(\frac{1}{5}\right)^2 = \frac{1}{5^2} = \frac{1}{25}$$
$$\left(\frac{2}{5}\right)^{-2} = \left(\frac{5}{2}\right)^2 = \frac{5^2}{2^2} = \frac{25}{4} = 6\frac{1}{4}$$

EXPONENTS OF EXPONENTS

- To take the exponent of an exponent, multiply the exponents.

$$(x^a)^b = x^{a \cdot b}$$

Examples:
$$(2^3)^5 = 2^3 \cdot 2^3 \cdot 2^3 \cdot 2^3 \cdot 2^3 = 2^{15}$$
$$(3^2)^4 = 3^8$$
$$\left(\left(\frac{3}{2}\right)^{-2}\right)^3 = \left(\frac{3}{2}\right)^{-6} = \left(\frac{2}{3}\right)^6 = \frac{2^6}{3^6} = \frac{64}{729}$$

COMMON ERRORS

$$7^2 + 7^4 \ne 7^6 \qquad\qquad 5^2 \cdot 7^2 \ne 12^2$$
$$7 \cdot 7 + 7 \cdot 7 \cdot 7 \cdot 7 \ne 7^6 \qquad 5 \cdot 5 \cdot 7 \cdot 7 \ne 12 \cdot 12$$

Radicals

- The square root of a positive number x is the number that when multiplied by itself equals x.

$$\sqrt{16} = \sqrt{4 \cdot 4} = \sqrt{4^2} = 4$$

- Equations in the form of $x^2 = positive\ constant$ have two solutions (two roots).

If $x^2 = 49$, then $x = 7$ or $x = -7$ (or $x = \pm 7$).

- The square root of a positive number is only the positive root.

$$\sqrt{16} = 4$$
$$\sqrt{16} \ne -4$$

SIMPLIFYING RADICALS

- If the number under the square root is a multiple of a perfect square, it can be simplified. First find the prime factorization of the number under the square root. Then look for pairs of the same prime number (perfect squares), and take one of them out of the square root (because $\sqrt{3 \cdot 3} = 3$).

  Examples: $\sqrt{75} = \sqrt{3 \cdot \underbrace{5 \cdot 5}} = 5 \cdot \sqrt{3} = 5\sqrt{3}$

  $\sqrt{108} = \sqrt{3 \cdot \underbrace{2 \cdot 2} \cdot \underbrace{3 \cdot 3}} = 2 \cdot 3 \cdot \sqrt{3} = 6\sqrt{3}$

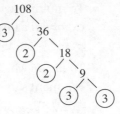

ADDITION AND SUBTRACTION OF RADICALS

- If the terms have the same number under the square root, they are considered like terms. Therefore, you can add or subtract them.

  Example: $3\sqrt{5} + 7\sqrt{5} - 13\sqrt{5} = (3 + 7 - 13)\sqrt{5} = -3\sqrt{5}$

- If the terms have different numbers under the square root, you cannot add or subtract them.

  Example: $8\sqrt{5} + 7\sqrt{3}$ does not simplify any further.

  Example: $\sqrt{45} + 17\sqrt{3} - 7\sqrt{5} - 6\sqrt{27}$ First simplify the radicals as much as you can.

  $\sqrt{3 \cdot 3 \cdot 5} + 17\sqrt{3} - 7\sqrt{5} - 6\sqrt{3 \cdot 3 \cdot 3}$

  $3\sqrt{5} + 17\sqrt{3} - 7\sqrt{5} - 3 \cdot 6\sqrt{3}$ Add the like terms.

  $3\sqrt{5} - 7\sqrt{5} + 17\sqrt{3} - 18\sqrt{3}$

  $-4\sqrt{5} - \sqrt{3}$

MULTIPLICATION AND DIVISION OF RADICALS

- To multiply two radicals, multiply the numbers under the square root sign.
  $\sqrt{x} \cdot \sqrt{y} = \sqrt{x \cdot y}$

- To divide two radicals, divide the numbers under the square root sign.
  $\dfrac{\sqrt{x}}{\sqrt{y}} = \sqrt{\dfrac{x}{y}}$

  Examples:

  $\sqrt{3} \cdot \sqrt{12} = \sqrt{3 \cdot 12} = \sqrt{36} = 6$

  $2\sqrt{5} \cdot 4\sqrt{7} = 8 \cdot \sqrt{35} = 8\sqrt{35}$

  $\dfrac{\sqrt{63}}{\sqrt{7}} = \sqrt{\dfrac{63}{7}} = \sqrt{9} = 3$

  $(4\sqrt{3})^2 = 4\sqrt{3} \cdot 4\sqrt{3} = 4 \cdot 4 \cdot \sqrt{3} \cdot \sqrt{3} = 16 \cdot 3 = 48$

  $\dfrac{3\sqrt{200}}{\sqrt{40}} = 3 \cdot \sqrt{\dfrac{200}{40}} = 3 \cdot \sqrt{5} = 3\sqrt{5}$

  $(2\sqrt{3} + 1)(2\sqrt{3} - 1) = 4\sqrt{9} - 2\sqrt{3} + 2\sqrt{3} - 1 = 12 - 1 = 11$

  (See the "Difference of Two Squares" on page 122 for a quicker solution.)

Although $(-4)^2 = 4^2 = 16$, $\sqrt{16}$ equals only 4.

However, when $x^2 = 49$, $x = 7$ or $x = -7$.

- Radicals can also be represented as fractional exponents (exponents in the form of a fraction).

$$\sqrt[b]{x^a} = x^{\frac{a}{b}}$$

$$x^{\frac{1}{a}} = \sqrt[a]{x}$$

$$\sqrt{5^3} = 5^{\frac{3}{2}}$$

$$3^{\frac{1}{3}} = \sqrt[3]{3}$$

NOTE

The square root of a number is that number raised to the power of $\frac{1}{2}$.

$$\sqrt{6} = (6)^{\frac{1}{2}}$$

COMMON ERRORS

$$\sqrt{x^2 + y^2} \neq x + y$$

$$\sqrt{25 + 16} \neq \sqrt{25} + \sqrt{16}$$

$$(x^4 - y^6)^{\frac{1}{2}} = \sqrt{(x^4 - y^6)} \neq x^2 - y^3$$

SAMPLE PROBLEMS

EXAMPLE 1

$$(5 - 5^{-1}) \cdot \frac{8 + 8^{-1}}{2^{-2} + 2^{-3}} = ?$$

$(5 - 5^{-1}) \cdot \dfrac{8 + 8^{-1}}{2^{-2} + 2^{-3}}$ Work out the exponents first. Remember $x^{-1} = \frac{1}{x}$.

$\left(5 - \dfrac{1}{5}\right) \cdot \dfrac{8 + \frac{1}{8}}{\frac{1}{2^2} + \frac{1}{2^3}}$ To add the fractions, first make the denominators equal.

$\left(\dfrac{25}{5} - \dfrac{1}{5}\right) \cdot \dfrac{\frac{64}{8} + \frac{1}{8}}{\frac{2}{8} + \frac{1}{8}}$ Add and subtract the fractions.

$\left(\dfrac{24}{5}\right) \cdot \dfrac{\frac{65}{8}}{\frac{3}{8}}$ Divide the fractions. Flip the fraction in the denominator and multiply it by the fraction in the numerator.

$\overset{8}{\underset{1}{\cancel{\dfrac{24}{5}}}} \cdot \overset{13}{\underset{1}{\cancel{\dfrac{65}{8}}}} \cdot \overset{1}{\underset{1}{\cancel{\dfrac{8}{3}}}} = 104$

EXAMPLE 2

$$\frac{7^{x+1} + 7^x}{7^x + 7^{x-1}} = ?$$

$\dfrac{7^{x+1} + 7^x}{7^x + 7^{x-1}} =$	Split the exponents using the rule $x^{a+b} = x^a \cdot x^a$.
$\dfrac{7^1 \cdot 7^x + 7^x}{7^x + 7^{-1} \cdot 7^x}$	Factor out 7^x from both the numerator and the denominator.
$\dfrac{7^x(7+1)}{7^x(1 + 7^{-1})}$	Simplify 7^x, and replace 7^{-1} with $\dfrac{1}{7}$.
$\dfrac{8}{1 + \dfrac{1}{7}}$	Make the denominators equal in order to add the fractions.
$\dfrac{8}{\dfrac{8}{7}} = 8 \cdot \dfrac{7}{8} = 7$	Flip the fraction in the denominator, and multiply.

EXAMPLE 3

What is the value of x if $8^x \cdot 16^{x+1} = 4^{x+7}$?

(A) 1

(B) $\dfrac{3}{2}$

(C) 2

(D) $\dfrac{5}{2}$

(E) 3

To multiply exponential functions, the bases need to be equal.

$8^x \cdot 16^{x+1} = 4^{x+7}$	Replace $8 = 2^3$, $16 = 2^4$, and $4 = 2^2$.
$(2^3)^x \cdot (2^4)^{x+1} = (2^2)^{x+7}$	To find the exponent of an exponent, multiply.
$2^{3x} \cdot 2^{4x+4} = 2^{2x+14}$	To multiply, add the exponents.
$2^{7x+4} = 2^{2x+14}$	If the bases are equal, the exponents must also be equal.
$7x + 4 = 2x + 14$	Subtract $2x$ from each side.
$5x + 4 = 14$	Subtract 4 from each side.
$5x = 10$	Divide by 5.
$x = 2$	

1. $\sqrt{64 - 36} = ?$

   (A) 1
   (B) $\sqrt{2}$
   (C) 2
   (D) $2\sqrt{7}$
   (E) 10

2. $x = -\dfrac{1}{(0.2)^2}$ $y = -\dfrac{1}{(0.3)^3}$ $z = -\dfrac{1}{0.01}$

   Which of the following must be true?

   (A) $z < y < x$
   (B) $z < x < y$
   (C) $x < y < z$
   (D) $x < z < y$
   (E) $y < x < z$

3. Which of the following must be positive if $zy^3 < 0$, $x^2y^3 > 0$, and $x^3z^2 < 0$?

   (A) x
   (B) y
   (C) z
   (D) xy
   (E) zy

4. $(\sqrt{5} - 2) \cdot (\sqrt{5} + 2) = ?$

   (A) 1
   (B) 2
   (C) $2\sqrt{5}$
   (D) 3
   (E) $4\sqrt{5}$

5. If $m^k = 64$, which of the following is not a possible value of $m - k$?

   (A) -4
   (B) 1
   (C) 6
   (D) 12
   (E) 63

6. $\sqrt{\dfrac{3}{16}} + \dfrac{\sqrt{27}}{4} = ?$

   (A) $\dfrac{\sqrt{3}}{4}$
   (B) $\dfrac{\sqrt{3}}{2}$
   (C) $\sqrt{3}$
   (D) $3\sqrt{3}$
   (E) $4\sqrt{3}$

7. If $25^x \cdot 5^4 = 25^{3x-2}$ what is the value of 5^x?

    (A) 0

    (B) 1

    (C) 2

    (D) 25

    (E) 125

8. Which of the following is the best approximation of $\sqrt{\dfrac{1}{0.0065}}$?

    (A) $\dfrac{1}{8}$

    (B) $\dfrac{2}{25}$

    (C) 8

    (D) 12.5

    (E) 125

9. Which of the following is false?

    (A) $(3^4)^2 = 3^8$

    (B) $5^3 \cdot 5^2 = 5^5$

    (C) $\left(\dfrac{2}{3}\right)^{-2} = \dfrac{9}{4}$

    (D) $5^3 + 5^4 = 5^7$

    (E) $\dfrac{7^{-6}}{7^{-3}} = 7^{-3}$

10. $(5^{-1} + 2^{-1})^{-1} = ?$

    (A) -7

    (B) $\dfrac{7}{10}$

    (C) $\dfrac{10}{7}$

    (D) 7

    (E) 10

11. $\sqrt{\dfrac{5^3 + 5^3 + 5^3}{3^3 + 3^3 + 3^3 + 3^3 + 3^3}} = ?$

    (A) $\sqrt{\dfrac{15}{12}}$

    (B) $\dfrac{15}{12}$

    (C) $\dfrac{5^3}{3^3}$

    (D) $\dfrac{5^2}{3^2}$

    (E) $\dfrac{5}{3}$

12. $\dfrac{25^{3-3x}}{5^{5-6x}} = ?$

   (A) 5

   (B) 5^x

   (C) 5^{3x}

   (D) 5^{3x-2}

   (E) 5^{11-12x}

13. m and k are positive integers, and $m \neq k$. If $m^k = k^m$, what is the value of $m \cdot k$?

   (A) 4

   (B) 6

   (C) 8

   (D) 10

   (E) 12

14. 10^{-3} is greater than which of the following?

   (A) 10^{-4}

   (B) $\dfrac{1}{10^3}$

   (C) 10^{-2}

   (D) $\dfrac{1}{0.1}$

   (E) 0.001

15. $m^{-1} + n^{-1} = \dfrac{5}{12}$ and $m^{-1} - n^{-1} = \dfrac{1}{12}$. What is the value of $\dfrac{n}{m}$?

   (A) $\dfrac{3}{5}$

   (B) $\dfrac{2}{3}$

   (C) $\dfrac{5}{6}$

   (D) $\dfrac{3}{2}$

   (E) $\dfrac{5}{3}$

16. $\sqrt{3125}$ is how many times greater than $125\sqrt{125}$?

   (A) 5

   (B) 2.5

   (C) 1.25

   (D) 0.05

   (E) 0.04

1	2	3	4	5	6	7	8	9	10	11	12	13	14	15	16
D	A	B	A	D	C	D	D	D	C	E	A	C	A	D	E

1. **D** Remember that $\sqrt{64 - 36} \neq \sqrt{64} - \sqrt{36}$. Subtract first and then simplify.

   $\sqrt{64 - 36} = \sqrt{28} = \sqrt{4 \cdot 7} = 2\sqrt{7}$.

2. **A** $x = -\dfrac{1}{0.04} \quad y = -\dfrac{1}{0.027} \quad z = -\dfrac{1}{0.01}$

   The negative number with the smallest absolute value is the largest.

   Look for the fraction with the smallest absolute value, meaning the fraction with the largest denominator. x is the largest, and z is the smallest.

3. **B** $zy^3 < 0$ either z or y is negative.

   $x^2y^3 > 0$ x^2 is always positive, so y must be positive. This also tells us that in the first statement, z must be negative.

   $x^3z^2 < 0$ x is negative since z^2 is always positive.

4. **A** Remember that $\sqrt{5} \cdot \sqrt{5} = 5$.   Multiply each term.

   $(\sqrt{5} - 2)(\sqrt{5} + 2) = 5 + 2\sqrt{5} - 2\sqrt{5} - 4 = 1$

   Alternatively, use the difference of two squares.

   $(a - b)(a + b) = a^2 - b^2$
   $(\sqrt{5} - 2)(\sqrt{5} + 2) = \sqrt{5}^2 - 2^2 = 5 - 4 = 1$

5. **D** All possible values of m and k are:

   $8^2 = 64 \qquad m - k = 6$
   $4^3 = 64 \qquad m - k = 1$
   $2^6 = 64 \qquad m - k = -4$
   $64^1 = 64 \qquad m - k = 63$

   Therefore, $m - k$ cannot be 12.

6. **C** Simplify the first fraction:

   $\sqrt{\dfrac{3}{16}} = \dfrac{\sqrt{3}}{\sqrt{16}} = \dfrac{\sqrt{3}}{4}$

   Simplify the second fraction:

   $\dfrac{\sqrt{27}}{4} = \dfrac{\sqrt{3 \cdot 9}}{4} = \dfrac{3\sqrt{3}}{4}$

   Add the fractions:

   $\dfrac{\sqrt{3}}{4} + \dfrac{3\sqrt{3}}{4} = \dfrac{4\sqrt{3}}{4} = \sqrt{3}$

7. **D** First write each term as a power of 5. For example, replace 25 by 5^2.

$$25^x \cdot 5^4 = 25^{3x-2}$$
$$(5^2)^x \cdot 5^4 = (5^2)^{3x-2}$$
$$5^{2x} \cdot 5^4 = 5^{6x-4}$$
$$2x + 4 = 6x - 4$$
$$8 = 4x$$
$$x = 2$$
$$5^x = 5^2 = 25$$

8. **D** $\dfrac{1}{0.0065}$ can be written as $\dfrac{1}{0.0065} = \dfrac{1}{\dfrac{65}{10,000}} = \dfrac{10,000}{65}$. To approximate $\sqrt{\dfrac{10,000}{65}}$, replace 65 by 64 since $\sqrt{64} = 8$.

$$\sqrt{\frac{10,000}{64}} = \sqrt{\frac{100^2}{8^2}} = \frac{100}{8} = 12.5$$

9. **D**

(A) This is an exponent of an exponent, so multiply.

(B) These have the same bases. To multiply, add the exponents.

(C) $\left(\dfrac{2}{3}\right)^2 = \dfrac{3^2}{2^2} = \dfrac{9}{4}$

(D) $5^3 + 5^4 \neq 5^7$

These cannot be added without first finding 5^3 and 5^4.

(E) $\dfrac{7^{-6}}{7^{-3}} = 7^{-6-(-3)} = 7^{-3}$

These have the same base. To divide, subtract the exponents.

10. **C** $(5^{-1} + 2^{-1})^{-1} = \left(\dfrac{1}{5} + \dfrac{1}{2}\right)^{-1} = \left(\dfrac{1}{5} \cdot \dfrac{2}{2} + \dfrac{1}{2} \cdot \dfrac{5}{5}\right)^{-1} = \left(\dfrac{2}{10} + \dfrac{5}{10}\right)^{-1} = \left(\dfrac{7}{10}\right)^{-1} = \dfrac{10}{7}$

Note that $(5^{-1} + 2^{-1})^{-1} \neq 5^1 + 2^1$. You can distribute the power only when there is a multiplication or a division operation inside the parentheses. For example, $(5^{-1} \cdot 2^{-1})^{-1} = 5^1 \cdot 2^1 = 10$.

11. **E** $\sqrt{\dfrac{5^3 + 5^3 + 5^3}{3^3 + 3^3 + 3^3 + 3^3 + 3^3}} = \sqrt{\dfrac{3 \cdot 5^3}{5 \cdot 3^3}} = \sqrt{\dfrac{5^2}{3^2}} = \dfrac{5}{3}$

12. **A** Replace 25 by 5^2.

$$\frac{25^{3-3x}}{5^{5-6x}} = \frac{(5^2)^{3-3x}}{5^{5-6x}} = \frac{5^{6-6x}}{5^{5-6x}} = 5^{6-6x-(5-6x)}$$
$$5^{6-6x-5+6x} = 5^1 = 5$$

13. **C** The easiest way here is to try numbers until the equation is met: $m^k = k^m$ works only for $4^2 = 2^4$

$$m \cdot k = 8$$

14. **A** Rewrite 10^{-3} as $\frac{1}{10^3} = \frac{1}{1000}$. Convert all answer choices to fractions to see which one is less than $\frac{1}{1000}$;

(A) $10^{-4} = \frac{1}{10,000}$ is less than $\frac{1}{1000}$.

(B) $\frac{1}{10^3} = \frac{1}{1000}$ equal to $\frac{1}{1000}$

(C) $10^{-2} = \frac{1}{10^2} = \frac{1}{100}$ is greater than $\frac{1}{1000}$.

(D) $\frac{1}{0.1} = 10$ is greater than $\frac{1}{1000}$.

(E) $0.001 = \frac{1}{1000}$ is equal to $\frac{1}{1000}$.

15. **D** Replace $m^{-1} = \frac{1}{m}$ and $n^{-1} = \frac{1}{n}$.

Add the two equations side by side:

$$\frac{1}{m} + \frac{1}{n} = \frac{5}{12}$$
$$+ \frac{1}{m} - \frac{1}{n} = \frac{1}{12}$$
$$\overline{}$$
$$\frac{2}{m} = \frac{6}{12}$$
$$6m = 24$$
$$m = 4$$

Plug $m = 4$ into the first equation to get n:

$$\frac{1}{4} + \frac{1}{n} = \frac{5}{12}$$
$$\frac{1}{n} = \frac{5}{12} - \frac{1}{4} = \frac{5}{12} - \frac{3}{12} = \frac{2}{12}$$
$$\frac{1}{n} = \frac{1}{6}$$
$$n = 6$$
$$\frac{n}{m} = \frac{6}{4} = \frac{3}{2}$$

16. **E** The phrase "how many times greater" translates into division. Divide $\sqrt{3125}$ by $125\sqrt{125}$.

$$\frac{\sqrt{3125}}{125\sqrt{125}} = \frac{1}{125}\sqrt{\frac{3125}{125}} = \frac{1}{125} \cdot \sqrt{25} = \frac{5}{125} = \frac{1}{25} = \frac{4}{100} = 0.04$$

SECTION 3.4: QUADRATIC EQUATIONS, FUNCTIONS, AND SYMBOLISM

Quadratic Equations

- Quadratic equations have an x^2 term, which means they are second-degree equations.
- The standard form of a quadratic equation is $ax^2 + bx + c = 0$.
  a, b, and c are real numbers where $a \neq 0$.

  Examples: $3x^2 - 5x + 90 = 0$, $-4x^2 + 3 = 0$, $x^2 + x = -7$

SOLVING QUADRATIC EQUATIONS

- To solve a quadratic equation, take the following steps.

  1. Move all terms to the left side of the equation to get a zero on the right.
  2. Simplify the left side if necessary by combining like terms.
  3. Try to use factoring to solve the equation as explained below.
  4. If the equation does not factor easily, use the quadratic formula.

- To factor a quadratic equation in the form of $x^2 + bx + c = 0$ (where $a = 1$), follow these steps.

  1. After setting the equation equal to zero, find two numbers (n_1 and n_2) that add to b and multiply to c.

     $x^2 + (\text{Sum})x + (\text{Product}) = 0$

  2. If you can find such numbers, factor the equation as $(x + n_1) \cdot (x + n_2) = 0$.
  3. Finally, set each factor equal to zero, such as $(x + n_1) = 0$, $(x + n_2) = 0$. Then solve for x.

Example: Solve $x^2 - x - 6 = 0$

First, find two numbers where $n_1 \cdot n_2 = -6$ and $n_1 + n_2 = -1$.

$n_1 = -3$ and $n_2 = 2$ because $-3 \cdot 2 = -6$ and $-3 + 2 = -1$.

Next, factor $x^2 - x - 6 = (x - 3) \cdot (x + 2) = 0$.

Finally, solve for x. Since $x - 3 = 0$ and $x + 2 = 0$, $x = 3$ and $x = -2$.

Check: $(-2)^2 - (-2) - 6 = 4 + 2 - 6 = 0$ ✓
$(3)^2 - (3) - 6 = 9 - 3 - 6 = 0$ ✓

ZERO PRODUCT RULE

If the product of two terms is zero, at least one of them has to equal zero.
If $p \cdot q = 0$ then $p = 0$ or $q = 0$.

- If factoring is not possible or if you cannot see it quickly, use the quadratic formula.

$$x = \frac{-b \pm \sqrt{b^2 - 4ac}}{2a}$$

Example: $x^2 - x - 6 = 0$
$a = 1, b = -1,$ and $c = -6$

$$x = \frac{-(-1) \pm \sqrt{(-1)^2 - 4(1)(-6)}}{2 \cdot 1} = \frac{1 \pm \sqrt{1 + 24}}{2} = \frac{1 \pm 5}{2}$$

$\frac{1 + 5}{2} = 3$ and $\frac{1 - 5}{2} = -2$. The solutions set is $S = \{3, -2\}$.

- If $b^2 - 4ac$ is less than zero, the quadratic equation has no real solutions.

Example: $3x^2 - 2x + 9 = 0$
$b^2 - 4ac = (-2)^2 - 4(3)(9) = 4 - 108 = -104 < 0$

NOTE

Factoring will work for most of the equations you will encounter on the test. Using the quadratic formula is your back-up plan.

NOTE

All quadratic equations can be solved by using the quadratic formula. In some cases, applying the formula takes a bit longer, but it always gives a solution.

- If $b^2 - 4ac$ is equal to zero, the quadratic equation has a single solution. This is actually a double root where two solutions are equal to each other.

    Example: $x^2 - 6x + 9 = 0$
    $$b^2 - 4ac = (-6)^2 - 4(1)(9) = 36 - 36 = 0$$

- If $b^2 - 4ac$ is greater than zero, the quadratic equation has two distinct real solutions.

    Example: $x^2 - 6x - 9 = 0$
    $$b^2 - 4ac = (-6)^2 - 4(1)(-9) = 36 + 36 = 72$$

DIFFERENCE OF TWO SQUARES

- The difference of two squares can be factored as follows:

    $a^2 - b^2 = (a - b) \cdot (a + b)$

    Examples: $x^2 - 49 = (x - 7) \cdot (x + 7)$
    $16x^2 - 9b^2 = (4x - 3b) \cdot (4x + 3b)$

PERFECT SQUARES

- The square of a binomial in the form of $(a + b)$ can be quickly determined with "first term^2 + (2 × first term × second term) + second term^2."

    $(a + b)^2 = a^2 + 2ab + b^2$

    Example: $(3x + 4b)^2 = (3x)^2 + 2 \cdot 3x \cdot 4b + (4b)^2 = 9x^2 + 24xb + 16b^2$

- The square of a binomial in the form of $(a - b)$ can be quickly determined with "first term^2 − (2 × first term × second term) + second term^2."

    $(a - b)^2 = a^2 - 2ab + b^2$

    Example: $(5t - 4b)^2 = (5t)^2 - 2 \cdot 5t \cdot 4b + (4b)^2 = 25t^2 - 40tb + 16b^2$

FACTORS AND ROOTS OF QUADRATIC EQUATIONS

- If $(x - k)$ is a factor of $ax^2 + bx + c$, then $ax^2 + bx + c$ is divisible by $(x - k)$. In other words, there's another binomial in the form of $(ax - m)$ where $(x - k) \cdot (ax - m) = ax^2 + bx + c$.

    Example: $(x - 6)$ is a factor of $x^2 - 4x - 12$.
    So there is another binomial in the form $(x - k)$ where
    $x^2 - 4x - 12 = (x - 6)(x - k)$.
    In this case, $k = -2$ because $(-6) \cdot (-k)$ must equal -12.

- If $(x - k)$ is a factor of $ax^2 + bx + $ c, then substituting k makes both expressions zero.

    Example: $(x - 4)$ is a factor of $x^2 + 2x - 24$.

    So if you make $x - 4 = 0$ and solve for x, $x = 4$ is a root of $x^2 + 2x - 24$.
    (A root is a solution when a quadratic equation is set equal to zero.)
    $x^2 + 2x - 24 = (4)^2 + 2(4) - 24 = 16 + 8 - 24 = 0$

Functions

- A function can be thought of as an equation or a rule that tells you how to associate the elements in one set (the domain) with the elements in another set (the range). A function transforms one number into another.

- Functions are usually represented by a letter followed by the variable used in the expression.

  Examples: $f(x) = 4x^2 + 3$ $g(t) = t^3 - 3t + 2$

  Name of
  The function

  Name of
  the variable

  x
  (input) $\Longrightarrow$ $\boxed{f(x)}$ $\Longrightarrow$ y or $f(x)$
  (output)

- $f(3)$ indicates that 3 should be plugged in for each variable. If $f(x) = 4x^2 + x$, then $f(3) = 4(3)^2 + 3 = 36 + 3 = 39$.

- The **domain** of a function is the set of all the values for which the function is defined. The **range** of a function is the set of all values that are the output, or result, of applying the function.

  Example: $y = x^2 - 4$ is a function. If you input $x = 3$, you get an output of $y = 5$.

Symbolism

- Questions with symbols define specific rules and use arbitrary symbols to define certain operations. Since these are nonstandard functions, each question starts by defining the symbol.

  Example: If $☺x = 5x - 3$, what is the value of $☺7$?
  $☺7 = 5 \cdot (7) - 3 = 32$

  Example: (circle divided into quadrants: x | y / z | t) is defined as $x^2 \cdot t - y \cdot z$. What is the value of (circle divided into quadrants: 4 | 7 / 1 | 3)?

  This function tells you to take the upper left number, square it, and multiply by the lower right number. Then subtract the product of upper right and lower left numbers.

  $4^2 \cdot 3 - 7 \cdot 1 = 48 - 7 = 41$

Sequences

- A sequence is a series of ordered numbers following a rule. 2, 4, 8, 16, 32, 64, . . . is a sequence. Each number in the sequence is called a **term** of the sequence. The first number is the first term (a_1), the second number is the second term (a_2), and so on. The general term of a sequence is a_n.
- Sequences are represented either by a few of their terms or by the rule describing the terms.

  Example: $a_{n+1} = (a_n)^2 + 3$ means each term is found by squaring the previous term and adding 3. For example, if $a_1 = 4$, $a_2 = (a_1)^2 + 3 = (4)^2 + 3 = 19$
  $a_3 = (a_2)^2 + 3 = (19)^2 + 3 = 364$

ARITHMETIC SEQUENCES

- In an arithmetic sequence, each successive number differs by a fixed amount from the previous number. The difference of consecutive terms is constant.

  Examples: 72, 78, 84, 90, . . . $78 - 72 = 6$, so 6 is the **common difference** (d)
  2, -4, -10, -16, . . . $d = -4 - 2 = -6$

- The nth term of an arithmetic sequence can be found by $a_n = a_1 + (n - 1) \cdot d$, where d is the common difference and a_1 is the first term.

  Example: What is the 50th term in the series 5, 9, 13, 17, . . . ?

  $d = 9 - 5 = 4$
  $a_{50} = a_1 + 50 - 1 \cdot d = 5 + (50 - 1) \cdot 4 = 201$

- The sum of n terms of an arithmetic sequence is half the number of terms times the sum of the first term and the last term.

  Sum of n terms $= \dfrac{n}{2}(a_1 + a_n)$

  Example: What is the sum of all even numbers between 4 and 62, inclusive?

  There are 30 terms between 4 and 62 including 62 and 4 $\left(\dfrac{62 - 4}{2} + 1 = 30 \right)$. So $n = 30$.

  Sum $= \dfrac{30}{2}(4 + 62) = 990$

GEOMETRIC SEQUENCES

- In a geometric sequence, each successive number is a fixed multiple of the previous number. The ratio of consecutive terms is constant.

  Examples: 7, 21, 63, 189, . . . $\dfrac{21}{7} = 3$, so 3 is the **common ratio** (r)

  9, 3, 1, $\dfrac{1}{3}$, $\dfrac{1}{9}$, . . . $r = \dfrac{3}{9} = \dfrac{1}{3}$

- The nth term of an geometric sequence can be found by $a_n = a_1 \cdot r^{n-1}$ where r is the common ratio and a_1 is the first term.

  Example: What is the 12th term in the series 5, 15, 75, . . . ?

  $r = \dfrac{15}{5} = 3$

  $a_{12} = a_1 \cdot r^{12-1} = 5 \cdot 3^{12-1} = 5 \cdot 3^{11}$

REPEATING SEQUENCES

- In a repeating sequence, a fixed set of numbers repeats, such as
  $2, 3, 1, -1, 2, 3, 1, -1, 2, 3, \ldots$

  Example: In the repeating sequence below, what number is in the 953rd place?
  $\underline{5, 2, -2, -3, 1,}\ 5, 2, -2, -3, 1, 5, 2, -2, \ldots$

  5 numbers are repeating. Divide 953 by 5 and find the remainder, which is 3.
  The 3rd number in the sequence is also in the 953rd place: -2.

SAMPLE PROBLEMS

EXAMPLE 1

If $(x - 7)$ is a factor of $x^2 + tx - 14$, what is the value of t?

(A) -5 (B) -3 (C) 1   (D) 3   (E) 5

Solve $x - 7 = 0$ and find its root: $x = 7$. So $x = 7$ is also a root of $x^2 + tx - 14$ (i.e., makes it zero).

$7^2 + t(7) - 14 = 0$	Multiply.
$49 + 7t - 14 = 0$	Add the like terms.
$7t + 35 = 0$	Subtract 35 from both sides.
$7t = -35$	Divide both sides by 7.
$t = -5$	

The answer is (A).

EXAMPLE 2

If $x = 0.35$, $0.2x^2 - 5x + 0.7$ is how much greater than $0.2x^2 - 7x - 0.3$?

(A) 0   (B) 0.4   (C) 1   (D) 1.7   (E) 2.7

Since the question is "how much greater," subtract the second polynomial from the first one. Carry out the subtraction before plugging in.

$0.2x^2 - 5x + 0.7 - (0.2x^2 - 7x - 0.3)$	Distribute the negative sign.
$0.2x^2 - 5x + 0.7 - 0.2x^2 + 7x + 0.3$	Group the like terms.
$0.2x^2 - 0.2x^2 - 5x + 7x + 0.7 + 0.3$	Add and subtract like terms.
$2x + 1$	Plug in 0.35 for x.
$2(0.35) + 1 = 1.7$	

The answer is (D).

EXAMPLE 3

If $4a^2 - b^2 = 27$ and $2a - b = 3$, what is the value of $2a + b + 1$?

(A) 3

(B) 4

(C) 9

(D) 10

(E) It cannot be determined from the information given

NOTE

Whenever you see a difference of two squares in a problem, factor it!

Whenever you see a difference of two squares in a problem, you should factor it.

$4a^2 - b^2 = (2a - b) \cdot (2a + b) = 27$ Since you know that $2a - b = 3$, plug that in.

$3 \cdot (2a + b) = 27$ Divide both sides by 3.

$(2a + b) = 9$ Add 1 to both sides.

$2a + b + 1 = 9 + 1$

$2a + b + 1 = 10$

The answer is (D).

EXAMPLE 4

If $a \,@\, b = 2a^2 + b$ and $a \,\#\, b = a \cdot b^2$, what is the value of $3 \,@\, (5 \,\#\, 2)$?

(A) 118

(B) 56

(C) 38

(D) 28

(E) 14

Evaluate the contents of the parentheses first. Since $a \,\#\, b = a \cdot b^2$, $5 \,\#\, 2 = 5 \cdot 2^2 = 5 \cdot 4 = 20$.

The question reduces to $3 \,@\, 20$. Since $a \,@\, b = 2a^2 + b$, $3 \,@\, 20 = 2 \cdot 3^2 + 20 = 2 \cdot 9 + 20 = 38$.

The answer is (C).

SECTION 3.4—PRACTICE PROBLEMS

1. $x = \frac{1}{2}$ is one of the solutions of $6x^2 + mx - 4 = -2$. What is the value of m?

(A) -1

(B) 0

(C) $\frac{1}{2}$

(D) 1

(E) 2

2. $\dfrac{64s^2 - 1}{8s + 1} = ?$

   (A) $-8s$

   (B) $8s - 1$

   (C) $1 - 8s$

   (D) $8s + 1$

   (E) $8s$

3. $x^2 + x - 5a + 3 = 0$. a is one of the roots of the equation. a could be which of the following?

   (A) -2

   (B) -3

   (C) 0

   (D) 2

   (E) 3

4. If $x \cdot (x - 12) = -35$, what is the difference of the values of x that satisfy the equation?

   (A) 2

   (B) 5

   (C) 7

   (D) 12

   (E) 22

5. $\dfrac{19^2 + 17 \cdot 19 + 19}{65^2 - 46^2} = ?$

   (A) 9

   (B) 3

   (C) 1

   (D) $\dfrac{1}{3}$

   (E) $\dfrac{1}{9}$

6. For what integer value of x is $2x^2 - 5x = -3$?

   (A) $\dfrac{3}{2}$

   (B) 1

   (C) 0

   (D) -1

   (E) -2

7. $\sqrt{205^2 - 84^2} = ?$

   (A) 289

   (B) 187

   (C) 153

   (D) 17

   (E) 9

8. If $x^2 + mx - 15$ is divisible by $(x - 5)$, what is the value of m?

   (A) 5
   (B) 2
   (C) −2
   (D) −3
   (E) −5

9. Which one of the following is a root of $\frac{1}{2x^2} - \frac{3}{4x} + \frac{1}{4} = 0$?

   (A) $-\frac{1}{2}$
   (B) $\frac{1}{2}$
   (C) $\frac{3}{2}$
   (D) 2
   (E) 3

10. If $4c^2 - d^2 = 19$ and $2c + d = 19$, what is the value of $d + c$?

   (A) 1
   (B) 5
   (C) 9
   (D) 14
   (E) 20

11. What is the product of two numbers that satisfy the equation $\sqrt{x + 3} - x = 3$?

   (A) −6
   (B) −2
   (C) 0
   (D) 3
   (E) 6

12. If one of the roots of $x^2 - (2m - 1)x + 3m + 2 = 0$ is equal to 1, what is the other root?

   (A) 10
   (B) 4
   (C) −1
   (D) −10
   (E) −13

13. If $x - 2y = 8$ and $xy = 5$, what is the value of $x^2 + 4y^2$?

   (A) 44
   (B) 54
   (C) 64
   (D) 74
   (E) 84

14. What is the value of $52^2 - 48^2 - 19^2$?

    (A) 33
    (B) 35
    (C) 37
    (D) 39
    (E) 41

15. If $x^2 + \dfrac{9}{x^2} = 31$, what is the value of $x - \dfrac{3}{x}$?

    (A) 36
    (B) 25
    (C) 9
    (D) 5
    (E) 3

16. For any integer greater than 1, «n» is defined as the remainder when n is divided by 5. If «n» = 3, what is the value of «$3n$»?

    (A) 1
    (B) 2
    (C) 3
    (D) 4
    (E) 9

Answer questions 17 and 18 based on the figure below.

For any integer, 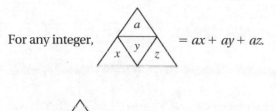 $= ax + ay + az.$

17. 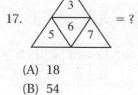 $= ?$

    (A) 18
    (B) 54
    (C) 152
    (D) 213
    (E) 630

18. If = 0. What is the value of a?

    (A) -180

    (B) $-\dfrac{1}{180}$

    (C) 0

    (D) $\dfrac{1}{180}$

    (E) 180

19. For all numbers x and y, the operation $\uparrow\downarrow$ is defined by $x\uparrow y\downarrow = \dfrac{x^2}{y^2}$. For all numbers k and t, the operation ∇ is defined by $k\nabla t = k^2 - t^2$. What is the value of $(9\nabla 7)\uparrow 64\downarrow$?

    (A) $\dfrac{1}{32}$

    (B) $\dfrac{1}{4}$

    (C) $\dfrac{49}{81}$

    (D) $\dfrac{81}{49}$

    (E) 4

20. For all numbers p and s, the operation $\exists$ is defined by $p\exists s = p^2 - 2p + ps$. If $3\exists s = 18$, what is the value of s?

    (A) 2

    (B) 5

    (C) 6

    (D) 57

    (E) 342

SECTION 3.4—SOLUTIONS

1	2	3	4	5	6	7	8	9	10	11	12	13	14	15	16	17	18	19	20
D	B	E	A	D	B	B	C	D	D	E	D	E	D	D	D	B	C	B	B

1. **D** You can find the value of m by substituting $x = \frac{1}{2}$.

$$6\left(\tfrac{1}{2}\right)^2 + m\left(\tfrac{1}{2}\right) - 4 = -2 \qquad \text{Add 4.}$$

$$6\cdot\tfrac{1}{4} + \tfrac{m}{2} = 2 \qquad \text{Simplify.}$$

$$\tfrac{3}{2} + \tfrac{m}{2} = 2 \qquad \text{Multiply both sides by 2.}$$

$$3 + m = 4 \qquad \text{Subtract 3 from both sides.}$$

$$m = 1$$

2. **B** $64s^2 - 1$ is a difference of two squares, $(8s)^2 - 1^2$. Factor it as $64s^2 - 1 = (8s - 1)(8s + 1)$.

$$\frac{(8s - 1)(8s + 1)}{8s + 1} = 8s - 1$$

3. **E** If a is one of the roots, you can plug it in for x and solve.

$a^2 + a - 5a + 3 = a^2 - 4a + 3 = 0$, which is another quadratic equation.

$$
\begin{aligned}
a^2 - 4a + 3 &= 0 \qquad &&\text{Factor (sum} = -4\text{, product} = 3\text{).}\\
(a - 1)(a - 3) &= 0\\
a - 1 = 0 \quad & a - 3 = 0\\
a = 1 \quad & a = 3
\end{aligned}
$$

4. **A**
$$
\begin{aligned}
x \cdot (x - 12) &= -35 \qquad &&\text{Distribute } x.\\
x^2 - 12x &= -35 &&\text{Add 35.}\\
x^2 - 12x + 35 &= 0 &&\text{Factor (sum} = -12\text{, product} = 35\text{).}\\
(x - 5)(x - 7) &= 0\\
x - 5 = 0 \quad & x - 7 = 0\\
x = 5 \quad & x = 7
\end{aligned}
$$

The difference is $7 - 5 = 2$.

5. **D** Since you can't use a calculator, use factoring to simplify your calculations.

Factor the numerator.

$19^2 + 17 \cdot 19 + 19 \qquad$ Factor out a 19.

$19(19 + 17 + 1) = 19 \cdot 37 \qquad$ Do not multiply yet!

Factor the denominator. $65^2 - 46^2$ is a difference of two squares. Write it as $(65 - 46)(65 + 46) = 19 \cdot 111$.

$$\frac{19 \cdot 37}{19 \cdot 111} = \frac{1}{3}$$

6. **B** This is a good example of a problem where you can plug in the answer choices for this problem.

Alternatively,

$$
\begin{aligned}
2x^2 - 5x &= -3 \qquad &&\text{Add 3 to both sides.}\\
2x^2 - 5x + 3 &= 0 &&\text{Use the quadratic formula.}
\end{aligned}
$$

$$\frac{-b \pm \sqrt{b^2 - 4ac}}{2a} = \frac{5 \pm \sqrt{25 - 4(2)(3)}}{2(2)} = \frac{5 \pm \sqrt{25 - 24}}{4} = \frac{5 \pm 1}{4}$$

$\dfrac{5 + 1}{4} = \dfrac{6}{4} = \dfrac{3}{2}$, which is not an integer.

$\dfrac{5 - 1}{4} = \dfrac{4}{4} = 1$

7. **B** Remember $\sqrt{205^2 - 84^2} \neq 205 - 84$. You need to square first, subtract, and then take the square root.

A quicker way is to factor using the difference of two squares.

$$\sqrt{205^2 - 84^2} = \sqrt{(205 + 84)(205 - 84)} = \sqrt{289 \cdot 121} = \sqrt{289} \cdot \sqrt{121} = 17 \cdot 11 = 187$$

8. **C** If $x^2 + mx - 15$ is divisible by $(x - 5)$, they have the same root.

Set $x - 5 = 0$ and find its root, $x = 5$. Substitute 5 for x in the first equation to find m.

$$5^2 + 5m - 15 = 0$$
$$25 + 5m - 15 = 0$$
$$5m + 10 = 0$$
$$5m = -10$$
$$m = -2$$

Alternatively, you can plug in the answer choices for this problem.

9. **D** This is one of the problems where you can plug in the answer choices. Try each one until you find a number that works. The challenge is that the answer choices are fractions. Plugging in might be very time consuming . To solve it algebraically, multiply both sides of the equation by $4x^2$, thereby simplifying the denominators. $4x^2$ is the least common multiple of all denominators.

$$4x^2\left(\frac{1}{2x^2} - \frac{3}{4x} + \frac{1}{4}\right) = (0)4x^2 \qquad \text{Multiply and simplify.}$$
$$\left(4x^2 \cdot \frac{1}{2x^2} - 4x^2 \cdot \frac{3}{4x} + 4x^2 \cdot \frac{1}{4}\right) = 0$$
$$2 - 3x + x^2 = 0 \qquad \text{Rearrange into standard form.}$$
$$x^2 - 3x + 2 = 0 \qquad \text{Factor (sum} = -3, \text{ product} = 2).$$
$$(x - 2)(x - 1) = 0$$
$$x - 2 = 0 \quad x - 1 = 0$$
$$x = 2 \qquad x = 1$$

10. **D** First factor $4c^2 - d^2$ as $(2c - d)(2c + d)$.

$$(2c - d)(2c + d) = 19 \qquad \text{You are given } 2c + d = 19.$$
$$(2c - d) \cdot 19 = 19$$
$$2c - d = 1$$

Solve the equations by elimination:

$$\begin{array}{r} 2c - d = 1 \\ + \ 2c + d = 19 \\ \hline 4c = 20 \\ c = 5 \end{array}$$

Plug $c = 5$ into one of the equations to find d.

$$2c + d = 19$$
$$2 \cdot 5 + d = 19$$
$$10 + d = 19$$
$$d = 9$$
$$d + c = 9 + 5 = 14$$

11. **E** $\sqrt{x+3} - x = 3$ Add x to both sides so that the square root is isolated.

$\qquad \sqrt{x+3} = 3 + x$ Square both sides.

$\qquad (\sqrt{x+3})^2 = (3+x)^2$

$\qquad\qquad x + 3 = 9 + 6x + x^2$ Subtract x and 3

$\qquad\qquad\quad 0 = 6 + 5x + x^2$ Rewrite.

$\qquad x^2 + 5x + 6 = 0$ Factor (sum = 5, product = 6).

$\qquad (x+2)(x+3) = 0$

$\qquad x = -2 \qquad x = -3$ Since any number under a square root must be positive or zero, make sure that you check the zeros you find by plugging them into the original equation. In this case, both -2 and -3 work.

Therefore, $(-2)(-3) = 6$

12. **D** Find the value of m by substituting 1 for x.

$\qquad 1^2 - (2m - 1) \cdot 1 + 3m + 2 = 0$

$\qquad\quad 1 - 2m + 1 + 3m + 2 = 0$

$\qquad m + 4 = 0 \qquad\qquad m = -4$

Therefore, the original equation is:

$\qquad x^2 - (2(-4) - 1)x + 3(-4) + 2 = 0$

$\qquad\qquad\qquad x^2 + 9x - 10 = 0$ Factor (sum = 9, product = -10).

$\qquad\qquad\quad (x + 10)(x - 1) = 0$

$\qquad\qquad x + 10 = 0 \qquad x - 1 = 0$

$\qquad\qquad\qquad x = -10 \qquad x = 1$

13. **E** The question is asking for the sum of the squares of two terms: x and $2y$.

It is possible and quicker to get to $x^2 + 4y^2$ without finding x and y separately. Square both sides of $x - 2y = 8$.

$\qquad\quad (x - 2y)^2 = 64$

$\qquad x^2 - 4xy + 4y^2 = 64$ $x^2 + 4y^2$ appears on the left.

$\qquad x^2 + 4y^2 - 4xy = 64$ You are given that $xy = 5$, so $4xy = 20$.

$\qquad\quad x^2 + 4y^2 - 20 = 64$

$\qquad\qquad x^2 + 4y^2 = 84$

14. **D** You can square each term and subtract. Alternatively, for the first part, $52^2 - 48^2$, use the difference of two squares.

$52^2 - 48^2 = (52 + 48)(52 - 48) = 100 \cdot 4 = 400$

So the question reduces to $400 - 19^2 = 400 - 361 = 39$.

Or, $400 - 19^2 = 20^2 - 19^2 = (20 - 19)(20 + 19) = 1 \cdot 39 = 39$

15. **D** Notice that if you square $x - \frac{3}{x}$, you get very similar terms to the first expression. You can find the square of the original expression and take the square root at the end.

$$\left(x - \frac{3}{x}\right)^2 = x^2 - 6 + \frac{9}{x^2} = x^2 + \frac{9}{x^2} - 6 \qquad x^2 + \frac{9}{x^2} = 31 \text{ is given.}$$

$$\left(x - \frac{3}{x}\right)^2 = 31 - 6 = 25$$

$$x - \frac{3}{x} = \pm 5$$

16. **D** «n» = 3 means when n is divided by 5, the remainder is 3. Pick a number for n that works for this case. For example, when 8 is divided by 5, the reminder is 3. Use $n = 8$ to answer the question. $3n = 3 \cdot 8 = 24$. When 24 is divided by 5, the remainder is 4.

17. **B** $ax + ay = az$ can be written as $a(x + y + z)$ by using factoring.

$$a(x + y + z) = 3(5 + 6 + 7) = 3 \cdot 18 = 54$$

18. **C** If $a \cdot (49 + 90 + 48) = 0$ then a must equal zero because the value inside the parentheses is not zero.

19. **B** Start by calculating the parentheses $(9 \, \nabla \, 7) = 9^2 - 7^2 = (9 - 7)(9 + 7) = 2 \cdot 16 = 32$. The question reduces to $32 \uparrow 64 \downarrow = \frac{32^2}{64^2} = \frac{32 \cdot 32}{64 \cdot 64} = \frac{1 \cdot 1}{2 \cdot 2} = \frac{1}{4}$.

20. **B** If $3 \exists s = 18$, then all p's get replaced by 3. $3 \exists s = 3^2 - 2 \cdot 3 + 3s = 18$.

$$9 - 6 + 3s = 18$$
$$3 + 3s = 18$$
$$3s = 15$$
$$s = 5$$

Word Problems

4

→ **4.1 TRANSLATION FROM WORDS TO EQUATIONS AND BASIC WORD PROBLEMS**

→ **4.2 RATE PROBLEMS**

→ **4.3 WORK PROBLEMS**

→ **4.4 MIXTURE PROBLEMS**

→ **4.5 INVESTMENT/INTEREST PROBLEMS**

→ **4.6 SET PROBLEMS AND VENN DIAGRAMS**

SECTION 4.1: TRANSLATION FROM WORDS TO EQUATIONS AND BASIC WORD PROBLEMS

Many students find it hard to comprehend or interpret word problems. Luckily, you can use a tool to unlock most word problems without memorizing specific formulas or solution methods for each type of word problem. Students who prefer the latter method usually get discouraged easily or baffled quickly when they encounter a slight change in wording or a derivative of a standard problem.

The method explained here is commonly called translation. We will start with certain ground rules to translate or convert word problems into algebraic equations that can be solved easily. If you have skipped the previous chapter, which reviews necessary algebra concepts, go back now and review that chapter. Focus especially on the sections in which we discuss linear equations with one unknown and linear equations with two unknowns.

Basic Translation Concepts

Let's review the fundamental translation first. This is where most students have trouble. We will put it all together at the end of this section. Take a look at the following examples.

Sixteen more than a number

$$16 \quad + \quad x = 16 + x$$

Eight more than five times a number

$$8 \quad + \quad 5 \cdot n = 8 + 5n$$

Eight less than six times a number

$$6 \cdot n \quad - \quad 8 \quad = 6n - 8$$

Notice that "less than" switches the position of our numbers in the translation. When we say "eight less than a number," we mean 8 is subtracted from a number or that number is decreased by 8.

Six percent of sixteen

$$\frac{6}{100} \quad \cdot \quad 16 = \frac{6}{100} \cdot 16 \quad \text{or} \quad 0.06 \cdot 16$$

The following table will help you translate certain expressions into symbols/operations. By watching for these words in word problems, you can translate them into corresponding parts of equations.

TRANSLATION TABLE

Words	Symbols	Examples
A number, which, what	$n, x, t \ldots$	Bob's age is: $B =$ a number is the same as: $n =$ what is: $x =$
Is, equals, was, has, will be, costs, the result is, is the same as	$=$	Two times a number is 6: $2 \cdot n = 6$ Jean's age was: $J =$
More than, plus, added to, combined, total	$+$	Five more than a number: $x + 5$
Less than, minus, subtracted from, decreased by, difference, the increase from	$-$	11 less than a number: $n - 11$ 7 is decreased by a number: $7 - a$ The difference between Ali's and Ken's ages: $A - K$ Five years ago, Jim's age: $j - 5$ The increase from ten to T: $T - 10$
Of, product of, times, multiplied by	$\cdot$	$\frac{1}{2}$ of a number: $\frac{1}{2} \cdot m$ or $0.5 \cdot m$ The product of a number and 9: $k \cdot 9$ or $9k$
Divided by, ratio of, per, out of	$/$ or $\div$	The ratio of 5 to 6: $\frac{5}{6}$ A number is divided by 12: $\frac{n}{12}$ or $n \div 12$
Percent, what percent	$\div 100$ $\frac{x}{100}$	12 percent: $\frac{12}{100}$ What percent of 60 is 15?: $\frac{x}{100} \cdot 60 = 15$
Consecutive integers	$n, n + 1,$ $n + 2, \ldots$	The sum of three consecutive integers equals: $x + (x + 1) + (x + 2) =$
Consecutive odd integers, consecutive even integers	$n, n + 2,$ $n + 4, \ldots$	The sum of two consecutive odd integers equals: $x + (x + 2) =$
Reciprocal of a number	$\frac{1}{d}$	A number plus twice its reciprocal: $d + 2 \cdot \frac{1}{d}$
Opposite of a number	$-n$	5 more than the opposite of a number: $5 + (-n)$ or $5 - n$

QUICK PRACTICE

Translate each of the following into an algebraic form. Do not solve.

1. Five more than the product of x and y

2. 25 is n less than 36

3. The ratio of x to y is the same as the ratio of z to 5

4. Kim's age 5 years from now

5. The sum of three consecutive even integers

6. The sum of twice a number and three times a number

7. Twice the sum of a number and three times the number

ANSWERS TO QUICK PRACTICE

1. $5 + xy$

2. $25 = 36 - n$

3. $\frac{x}{y} = \frac{z}{5}$

4. $K + 5$

5. $n + (n + 2) + (n + 4)$

6. $2x + 3x$

7. $2 \cdot (m + 3m)$

Translating Equations

Now let's proceed with translating full sentences, which produces equations.

Five times a number is twelve more than the number

$\quad\ 5 \quad\ \cdot \quad\ n \quad\ = \quad 12 \quad\ + \quad\ n \qquad\qquad 5n = 12 + n$

Since the second part states "the number," it refers to the same number defined in the first part of the statement. Therefore, we use the same letter (n) for both numbers.

What number equals twenty-two more than three times itself?

$\quad\ m \qquad\quad = \qquad 22 \qquad\quad + \qquad 3 \quad \cdot \quad m \qquad m = 22 + 3m$

The sum of two consecutive integers is 65.	$n + n + 1 = 65$
The ratio of $3x$ to $5y$ is equal to the ratio of 7 to 12.	$\frac{3x}{5y} = \frac{7}{12}$
What number equals sixteen less than twice itself?	$x = 2x - 16$
Seventy-five divided by a number is three times the number.	$\frac{75}{k} = 3k$
The sum of two consecutive even numbers is 30.	$z + z + 2 = 30$
The length of a rectangle is twice its width.	$l = 2w$
A triangle with sides a, $a + 2$, and 22 has a perimeter of 44	$a + a + 2 + 22 = 44$

Jason is eleven years older than Karen. $\qquad$ $J = 11 + K$

Five years ago, Jason was twice as old as Karen $\qquad$ $J - 5 = 2 \cdot (K - 5)$

In the last word problem, 5 years ago both Jason and Karen were 5 years younger than they are now. So we use $(J - 5)$ and $(K - 5)$ for their current ages.

AGE PROBLEMS

When time changes in age problems, everybody is affected. Two years ago means (current age − 2) for each person.

Two years from now, Tom will be three times as old as his brother $\qquad$ $T + 2 = 3 \cdot (B + 2)$

After Jeff gives \$30 to Kim, he will have half the money Kim has. $\qquad$ $J - 30 = \frac{1}{2}(K + 30)$

EXCHANGE PROBLEMS

When two parties exchange objects (or money), subtract the amount from the giver and add the same amount to the receiver.

Word Problems Solved by Translation

Many word problems will reveal themselves easily after a straightforward translation. We will cover such problems in this section and review other word problems that require simple formulas in the following sections of this chapter.

STEP-BY-STEP PLAN FOR SOLVING WORD PROBLEMS

Evaluate

STEP 1 Read the question very carefully to understand the problem in its entirety.

STEP 2 Identify the unknown quantities, and assign variables (a one-letter "name") to each quantity. Express each quantity/unknown in terms of one variable if possible.

Kyle's age: K
Ben is twice as old as Kyle: $2K$
The time Allen takes to finish the task: t
Barbara finishes the same task in 2 hours more than Allen: $t + 2$

NOTE

You should name the variables using easily identifiable letters (like the first letter of the unknown) rather than always calling them x or n.

Translate

STEP 3 Translate the sentences, which are essentially algebraic equations, into their mathematical equivalents.

STEP 4 Solve the algebraic equations to find the quantities you identified in Step 2.

Solve & Check

STEP 5 Go back to the question, read the last statement and make sure you answer the right question. Make sure your answer makes sense and that it is reasonable based on the facts presented in the question.

EXAMPLE 1

> Find the largest of three consecutive integers whose sum is 84.

We can start by translating parts of the question.

Let's name the three consecutive integers as n, $(n + 1)$, and $(n + 2)$.

The sum of three consecutive integers becomes $n + (n + 1) + (n + 2)$.

We complete the translation as $n + (n + 1) + (n + 2) = 84$.

$$3n + 3 = 84$$
$$3n = 81$$
$$n = 27$$
$$n + 1 = 28$$
$$n + 2 = 29$$

The largest consecutive integer is 29.

BE CAREFUL!

Always go back to the question and find out what it is asking once you've solved for the unknown.

Alternatively, questions about the sum of consecutive integers or consecutive odd/even integers can also be solved using the following shortcut.

SHORTCUT

The average of a set of consecutive integers is always the middle integer.

If the sum of a set of integers is given, divide the sum by the number of integers. The result is always the middle integer.

In the example above, $\frac{84}{3} = 28$ is the middle number. So the largest consecutive integer is 29.

- If an even number of integers is given, dividing the sum by the number of integers will give you a number between the two middle numbers.

For example, let's say the sum of six consecutive even integers is 102.

Divide 102 by 6, $\frac{102}{6} = 17$.

This means the middle two even numbers are 16 and 18. So the consecutive even integers are 12, 14, 16, 18, 20, and 22.

EXAMPLE 2

Katherine is 4 years older than Julia. 2 years from now, Julia's age will be $\frac{2}{3}$ of Katherine's age. How old was Katherine 5 years ago?

Let Katherine's age be K and Julia's age be J.

Today: $K = J + 4$

	Katherine	Julia
Today	$J + 4$	J
2 years from now	$J + 6$	$J + 2$

$J + 2 = \frac{2}{3}(J + 6)$	Julia's age will be $\frac{2}{3}$ of Katherine's age in 2 years.
	Multiply both sides by 3.
$3 \cdot (J + 2) = 2 \cdot (J + 6)$	Distribute.
$3J + 6 = 2J + 12$	Rearrange the terms.
$J = 6$	Julia is 6 years old today
$K = J + 4$	Katherine is 4 years older than Julia
$K = 6 + 4 = 10$	Katherine is 10 years old today.

Katherine was $10 - 5 = 5$ years old 5 years ago.

EXAMPLE 3

Sophie has twice as much money as Nate does. After Sophie gives $5 to Nate, she will have 1.5 times as much as Nate does. How much money did Nate have originally?

Let Sophie's age be S and Nate's age be N.

Originally: $S = 2 \cdot N$

	Sophie	Nate
Before	$2N$	N
After	$2N - 5$	$N + 5$

$2N - 5 = 1.5 \cdot (N + 5)$	Sophie's money is 1.5 times as much as Nate's.
$2N - 5 = 1.5N + 7.5$	Distribute 1.5.
$0.5N = 12.5$	Rearrange the terms.
$N = \$25$	Multiply both sides by 2.
	Nate has $25 today.

EXAMPLE 4

A triangle's sides are x, $x + 3$, and $x + 5$. If its perimeter equals 59, how long is the longest side?

Translate the "sum of" into addition.

$$x + (x + 3) + (x + 5) = 59$$
$$3x + 8 = 59$$
$$3x = 51$$
$$x = 17$$
$$x + 3 = 20$$
$$x + 5 = 22$$

However, the answer is not 17. Take a moment to go back to the question and double-check what is being asked.

The sides are 17, 20, and 22.

The longest side is 22.

SECTION 4.1—PRACTICE PROBLEMS

1. Sebastian was 6 years old 5 years ago. How old will he be 12 years from now?

   (A) 13
   (B) 17
   (C) 18
   (D) 23
   (E) 28

2. Hannah is four years older than her sister. Two years ago, her sister's age was $\frac{3}{4}$ of Hannah's age. How old will her sister be in 5 years?

   (A) 14
   (B) 18
   (C) 19
   (D) 21
   (E) 23

3. Sonny is 5 times as old as Michael. Three years from now, he will be twice as old as Michael. How old is Michael?

   (A) 1
   (B) 4
   (C) 5
   (D) 9
   (E) 11

4. Wendy is 16 years older than her brother Quentin. Three years from now, she will be 3 times as old as her brother. How old is Quentin?

(A) 24
(B) 21
(C) 8
(D) 5
(E) 3

5. The total price of a pair of shoes and a sweater is $365. When Julia exchanged the sweater for a cheaper one, the new total price became $293. If the price of the first sweater is 1.6 times the price of the second sweater, what is the price of the shoes?

(A) $72
(B) $120
(C) $173
(D) $192
(E) $245

6. 5 roommates share their rent equally. If one of the roommates moves out, they would each have to pay $260 more. How much is the total rent?

(A) $1,300
(B) $2,340
(C) $2,600
(D) $3,120
(E) $5,200

7. Joe and Tricia are waiting in a line. Tricia is ahead of Joe, and there are 5 people between them. If the sum of the number of people in front of Tricia and the total number of people in front of Joe is 22, how many people are in front of Tricia?

(A) 6
(B) 7
(C) 8
(D) 9
(E) 10

8. A bicycle's front tire has a diameter that is $\frac{1}{3}$ of the diameter of the back tire. If the back tire makes 90 revolutions, how many revolutions does the front tire make?

(A) 10
(B) 30
(C) 90
(D) 270
(E) 810

9. The largest of the 4 consecutive even integers is 150% of the smallest. What is the average of the two middle integers?

(A) 11
(B) 12
(C) 13
(D) 14
(E) 15

10. The square of an even number equals 6 less than 5 times the number. Which of the following is the square of the number?

(A) 2
(B) 3
(C) 4
(D) 6
(E) 9

11. The sum of the smallest and the largest of 7 consecutive integers is 128. What is the average of the 7 integers?

(A) 60
(B) 61
(C) 62
(D) 63
(E) 64

12. If the difference of two positive integers is 4 and the difference of their squares is 80, what is the sum of their squares?

(A) 8
(B) 12
(C) 20
(D) 144
(E) 208

13. A tank contains a certain amount of water. If x gallons of water are added, only $\frac{1}{4}$ of the tank would be empty. If x gallons are removed, only $\frac{1}{4}$ of the tank is full. What fraction of the tank is full?

(A) $\frac{1}{8}$

(B) $\frac{1}{4}$

(C) $\frac{3}{8}$

(D) $\frac{1}{2}$

(E) $\frac{3}{4}$

14. The cost of renting a villa in Puerto Vallarta is $\$P$, which is to be shared equally among 8 people. If two more people are invited to join, how much less will each of the original 8 people pay in terms of P?

(A) $\frac{P}{8}$

(B) $\frac{P}{10}$

(C) $\frac{P}{80}$

(D) $\frac{P}{40}$

(E) $\frac{P}{20}$

SECTION 4.1—SOLUTIONS

1	2	3	4	5	6	7	8	9	10	11	12	13	14
D	C	A	D	C	E	C	D	E	C	E	E	D	D

1. **D** If Sebastian was 6 years old 5 years ago, he is 11 now.

    12 years from now he will be $11 + 12 = 23$ years old

2. **C**

    Hannah : H
    Sister : S
    Today : $H = S + 4$

	Hannah	Sister
Today	$S + 4$	S
2 years ago	$S + 4 - 2$	$S - 2$

    $$S - 2 = \frac{3}{4}(S + 4 - 2)$$

$S - 2 = \frac{3}{4}(S + 2)$	Multiply both sides by 4.
$4(S - 2) = 3(S + 2)$	Distribute 4, and distribute 3.
$4S - 8 = 3S + 6$	Rearrange.
$S = 14$	Today
$S = 19$	5 years from now

3. **A**

    Sonny : S
    Michael : M
    Today : $S = 5M$

	Sonny	Michael
Today	$5M$	M
3 years from now	$5M + 3$	$M + 3$

$5m + 3 = 2(m + 3)$	Distribute 2.
$5m + 3 = 2m + 6$	Rearrange.
$3m = 3$	Divide by 3.
$m = 1$	Michael is 1 year old today.

4. **D**

Wendy : W

Quentin : Q

Today : $W = Q + 16$

	Wendy	**Quentin**
Today	$Q + 16$	Q
3 years from now	$Q + 16 + 3$	$Q + 3$

$Q + 19 = 3 \cdot (Q + 3)$ Distribute 3.

$Q + 19 = 3Q + 9$ Rearrange.

$\quad\ 10 = 2Q$ Divide by 2.

$\quad\ Q = 5$

5. **C**

Shoes + Sweater 1 = \$365

Shoes + Sweater 2 = \$293

Note that the difference between the two totals is also the difference between the prices of the two sweaters.

The price of the first sweater is 1.6 times the price of the second sweater.

$S_1 = 1.6 \cdot S_2$

The price difference between the sweaters is

\$365 − \$293 = \$72.

This means $S_1 - S_2 = 72$.

Replace S_1 by $1.6S_2$ and solve for S_2.

$1.6S_2 - S_2 = 72$

$\quad\ 0.6S_2 = 72$

$\quad S_2 = \dfrac{72}{0.6} = \$120$

If the second sweater is \$120, the shoes are (\$293 − \$120) = \$173.

6. **E** Let the total rent be R.

If 5 people are sharing the rent equally, each person pays $\dfrac{R}{5}$.

If 4 people are sharing the rent equally, each person pays $\dfrac{R}{4}$.

The increase in rent per person is the difference between the above quantities

$$\frac{R}{4} - \frac{R}{5} = 260$$

$$\frac{5R}{20} - \frac{4R}{20} = 260$$

$$\frac{R}{20} = 260$$

Rent $= 20 \cdot 260 = \$5,200$

7. **C** A simple diagram may help in this problem

J ☺☺☺☺ T $\underbrace{n \text{ people}}$ ⌂

Let the number of people in front of T be n.

The number of people in front of J becomes $n + 1 + 5$.

The number of people in front of J and T is $n + n + 1 + 5 = 22$.

$2n + 6 = 22$

$\quad 2n = 16$

$\quad\quad n = 8$

There are 8 people in front of Tricia.

8. **D** Clearly, both tires will travel the same distance. When a tire makes one revolution, it moves forward as far as its circumference.

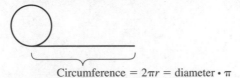

Circumference $= 2\pi r =$ diameter $\cdot \pi$

Diameter of the front tire $= d$

Diameter of the rear tire $= 3d$

After 90 revolutions, the back tire travels:

Circumference $\cdot$ revolutions $= \pi \cdot 3d \cdot 90$

The front tire will travel the same distance:

$d \cdot \pi \cdot$ front revolutions $= \pi \cdot 3d \cdot 90$

Front revolutions $= 3 \cdot 90 = 270$

9. **E** Let the 4 consecutive integers be

$n, (n + 2), (n + 4),$ and $(n + 6)$

The largest is 150% of the smallest translates into:

$n + 6 = 1.5n$

$\quad 6 = 0.5n$

$\quad n = 12$

So the numbers are 12, 14, 16, and 18.

The average of the two middle numbers is $\dfrac{14 + 16}{2} = 15$.

10. **C** We can translate the problem as follows

$\quad\quad\quad x^2 = 5x - 6$ $\quad$ Quadratic equation

$\quad x^2 - 5x + 6 = 0$ $\quad\quad$ Factor or use the quadratic formula

$(x - 3)(x - 2) = 0$

$x = 3$ or $x = 2$

Since we are told it is an even number, $x = 2$.

The answer is the square of $2^2 = 4$.

11. **E** Let the 7 consecutive integers be

$n, (n + 2), \ldots, (n + 6)$

The sum of the largest and smallest is 128.

$n + n + 6 = 128$

$2n = 122$

$n = 61$

So the numbers are 61, 62, 63, 64, 65, 66, and 67.

Their average would be the middle number, 64.

Alternatively, you could simply add them and divide by 7.

12. **E** The difference of two positive integers is 4 translates into $x - y = 4$.

The difference of their squares is 80 translates into $x^2 - y^2 = 80$.

Notice that the second equation is a difference of two squares and can be factored as follows.

$x^2 - y^2 = (x - y)(x + y) = 80$

From the first equation, we know that $x - y = 4$. So we can plug it in.

$4 \cdot (x + y) = 80$

$x + y = 20$ ___ Now we have two first-degree equations that we can solve together.

$x + y = 20$

$\underline{x - y = 4}$ ___ Recall the first equation here.

Add the two equations side by side.

$2x = 24$ ___ Divide both sides by two.

$x = 12$

$y = 12 - 4 = 8$

$x^2 + y^2 = 12^2 + 8^2 = 144 + 64 = 208$

13. **D** $\frac{1}{4}$ empty means $\frac{3}{4}$ full.

Let the initial amount of water be t.

$t + x = \frac{3}{4}$ of the tank

$+ \ t - x = \frac{1}{4}$ of the tank ___ We can add the equations side by side to eliminate x.

$\rule{4cm}{0.4pt}$

$2t = \frac{4}{4}$ of the tank

$t = \frac{1}{2}$ of the tank

Alternatively, draw a simple figure to help.

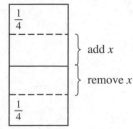

14. **D** The original rent per person is $\frac{P}{8}$.

The rent per person after 2 joined in is $\frac{P}{10}$.

$$\frac{P}{8} - \frac{P}{10} = \frac{5P}{40} - \frac{4P}{40} = \frac{P}{40}$$

SECTION 4.2: RATE PROBLEMS

The word "per" reminds us that we are working with a rate. This is simply a ratio of two quantities with different units. Thinking of rates as fractions may simplify our understanding of these types of problems. For example, a consultant's rate is the dollar amount she charges per hour, $/hr. Sales per month for a grocery store, miles per gallon for a car, or fee per mile for a cab are all everyday rates we are familiar with.

In general, any rate problem can be set up as follows:

$$\text{Rate (quantity } A \text{ per quantity } B) \cdot \text{Quantity } B = \text{Quantity } A$$

Miles per gallon times gallons used = Miles driven $\qquad \frac{\text{miles}}{\text{gallon}} \cdot \text{gallons} = \text{miles}$

Rate of service ($ per hour) times hours = Total fee $\qquad \frac{\$}{\text{hour}} \cdot \text{hours} = \$$

If a certain bookstore sells 127 books every two days, at this rate how many books will it sell in 14 days?

Books per day · Number of days = Total books sold

$$\frac{127}{2} \cdot 14 = 889 \text{ books}$$

Alternatively, set up a proportion.

$$\frac{127 \text{ books}}{2 \text{ day}} = \frac{x \text{ books}}{14 \text{ days}}$$

The fundamental relationship to solve rate problems involving motion is:

$$\text{Rate} \cdot \text{Time} = \text{Distance} \qquad \text{or} \qquad r \cdot t = d \qquad \frac{\text{miles}}{\text{hour}} \cdot \text{hours} = \text{miles}$$

In other words, distance traveled D by an object is the product of its average speed r and the total time t of motion. The average speed (rate) is the rate of change of distance or the distance per time.

Another easy way to remember this relationship is from the speed limit signs we use on the highway every day. Rate $= \frac{\text{mi}}{\text{hr}}$ (miles per hour).

So r must equal $\frac{\text{mi}}{\text{hr}}$, which is $\frac{\text{distance (mi)}}{\text{time (hr)}}$.

We can also rearrange the $r \cdot t = D$ formula to calculate time or rate.

$$D = r \cdot t$$
$$t = \frac{D}{r}$$
$$r = \frac{D}{t}$$

Notice how the units work. In our basic $r \cdot t = D$ formula, $\frac{\text{mi}}{\text{hr}} \cdot \text{hr} = \text{mi}$.

SAMPLE PROBLEMS

EXAMPLE 1

Helen bikes from home to work at an average speed of 15 miles per hour. On the way back, she runs at 9 miles per hour. Her run takes 20 minutes longer than her bike ride. How far is her work from home?

Notice that in this problem, the distance covered in both legs of the trip are the same. So $d_1 = d_2$.

Let the time Helen takes to bike to work be t

Since 20 minutes equals $\frac{1}{3}$ hours, the time Helen takes to run back becomes $t + \frac{1}{3}$.

	Rate ×	Time =	Distance
Bike	15	t	$15t$
Run	9	$t + \frac{1}{3}$	$9 \cdot \left(t + \frac{1}{3}\right)$

Because the distances are the same, set $d_1 = d_2$.

$$15t = 9 \cdot \left(t + \frac{1}{3}\right)$$ Distribute 9.

$$15t = 9t + 3$$ Subtract $9t$ from both sides.

$$6t = 3$$ Divide both sides by 3

$$t = \frac{1}{2} \text{ hrs}$$ Biking time is $\frac{1}{2}$ hour or 30 minutes.

$$D = r \times t = 15 \times \frac{1}{2} = 7.5 \text{ miles}$$ Use the main distance formula.

EXAMPLE 2

A car leaves City A and heads toward City B, which is 240 miles away, traveling at an average speed of 45 miles per hour. If another car leaves from City B and heads toward City A at the same time and travels at an average speed of 35 miles per hour, how far away from City A do they meet?

Notice that in this problem, the total distance is covered by two cars together. As a result, we will set up the final equation as $d_1 + d_2 = d_{total}$.

Since they travel until they meet, the time of travel for both cars is the same, t.

	Rate ×	Time =	Distance
Car from A	45	t	$45t$
Car from B	35	t	$35t$
Total			240

$$d_1 + d_2 = d_{total}$$
$$45t + 35t = 240$$
$$80t = 240$$
$$t = 3 \text{ hours}$$

Now check what the question is asking.

Distance from City A = Rate of Car A times time = $45 \cdot 3 = 135$ miles.

EXAMPLE 3

A boat travels 72 miles to an island at 24 miles per hour and returns at 12 miles per hour. What is its average speed for the entire trip?

$$t_1 = \frac{D}{r_1} = \frac{72}{24} = 3 \text{ hours}$$

$$t_2 = \frac{D}{r_2} = \frac{72}{12} = 6 \text{ hours}$$

$$\text{Average rate} = \frac{\text{Total distance traveled}}{\text{Total time}} = \frac{72 \cdot 2}{9} = 16 \text{ miles per hour}$$

SECTION 4.2—PRACTICE PROBLEMS

1. A bicycle left point A and traveled north at 12 km/hr. Two hours later, a motorcycle left the same point riding south at 34 km/hr. If at the end of t hours, they are 162 km apart, how long was the motorcycle's ride?

   (A) 7 hours
   (B) 6 hours
   (C) 5 hours
   (D) 4 hours
   (E) 3 hours

2. A car travels 240 miles from Town A to Town B at an average speed of 60 miles per hour. At what speed did it travel on the way back if its average speed for the whole trip was 48 miles per hour?

   (A) 16 miles per hour
   (B) 24 miles per hour
   (C) 36 miles per hour
   (D) 40 miles per hour
   (E) 48 miles per hour

3. Two cars leave City A and travel in opposite directions. The first car travels at 65 miles per hour, and the second car travels at 55 miles per hour. How many hours pass before they are 300 miles apart?

   (A) 2.0 hours
   (B) 2.5 hours
   (C) 3.0 hours
   (D) 3.5 hours
   (E) 4.0 hours

4. Karen drives the first 180 miles of her trip at an average speed of 60 miles per hour. If she drives the remaining 120 miles at an average speed of 30 miles per hour, the average speed for her entire trip is closest to which of the following?

   (A) 42 miles per hours
   (B) 43 miles per hours
   (C) 45 miles per hours
   (D) 47 miles per hours
   (E) 50 miles per hours

5. A company produces its cell phones in two different countries. The cost to manufacture one phone in country *A* is 20% more than it is in Country *B*. The company produces 60,000 phones in Country *A* and 48,000 phones in Country *B*. If the average cost per phone is $10, what is the cost per phone in Country *A*?

   (A) $9.00
   (B) $9.20
   (C) $9.80
   (D) $10.20
   (E) $10.80

6. Carl and Jen's houses are 15 miles apart. If Carl leaves his house biking 12 miles per hour toward Jen's house and Jen leaves her house at the same time running toward Carl's house at 8 miles per hour. How much time passes until they meet?

   (A) 1 hour 15 minutes
   (B) 1 hour
   (C) 45 minutes
   (D) 30 minutes
   (E) 15 minutes

7. A tortoise starts walking due north at a rate of 2 kilometers per hour. 2 hours later, a hare starts running in the same direction from the same point at a rate of 6 kilometers per hour. How long after the tortoise's departure will the hare catch up to the tortoise?

   (A) 1 hour
   (B) 1.5 hours
   (C) 2 hours
   (D) 2.5 hours
   (E) 3 hours

8. Mark charges *c* dollars per *h* hours of consulting services. At this rate, how much would it cost to hire Mark for a project that is expected to take him *k* hours?

   (A) $\dfrac{c \cdot h}{k}$

   (B) $\dfrac{c \cdot k}{h}$

   (C) $\dfrac{k \cdot h}{c}$

   (D) $\dfrac{c \cdot h \cdot k}{60}$

   (E) $\dfrac{60 \cdot c \cdot h}{k}$

9. At the same time, two cars take off from City A and travel toward City B with average speeds of 40 miles per hour and 50 miles per hour. The faster car reaches City B 15 minutes earlier than slower car. How far is City B from City A?

(A) 10 miles
(B) 40 miles
(C) 50 miles
(D) 60 miles
(E) 90 miles

10. Daria has k friends on the FaceSpace online community. If she adds friends at a constant rate of m friends per week for the next n days, how many friends will she have at the end of n days?

(A) kmn
(B) $k + nm$
(C) $\dfrac{nm}{7}$
(D) $k + \dfrac{nm}{7}$
(E) $k + n + m$

11. Camille biked to work at an average speed of 14 miles per hour and returned home using the same route at an average speed of 18 miles per hour. If the return trip took 20 minutes less than the first trip, how long was her total commute?

(A) 6 miles
(B) 12 miles
(C) 14 miles
(D) 21 miles
(E) 42 miles

12. On a 90-mile circular track, two cars start driving from the same point in opposite directions. Car A's average speed is 120 miles per hour, and Car B's average speed is 150 miles per hour. How many miles will Car A have traveled by the time they meet for the first time?

(A) 30 miles
(B) 35 miles
(C) 40 miles
(D) 45 miles
(E) 50 miles

SECTION 4.2—SOLUTIONS

1	2	3	4	5	6	7	8	9	10	11	12
E	D	B	B	E	C	E	B	C	D	E	C

1. **E**

	Rate	×	Time	=	Distance
Bicycle	12		t		$12t$
Motorcycle	34		$(t-2)$		$34(t-2)$
Total					162

The time traveled by the motorcycle is $(t-2)$ because it takes off two hours after the bicycle.

Since the total distance between them is given and they travel in exactly opposite directions, we can add each distance traveled to make it equal to 162.

$$d_1 + d_2 = d_{total}$$
$$12t + 34(t-2) = 162 \quad \text{Distribute 34.}$$
$$12t + 34t - 68 = 162 \quad \text{Rearrange.}$$
$$46t = 230 \quad \text{Divide by 46.}$$
$$t = 5$$
$$(t-2) = (5-2) = 3 \text{ hours}$$

The motorcycle traveled for 3 hours.

2. **D** For the first part of the trip:

$$t_1 = \frac{D}{r_1} = \frac{240}{60} = 4 \text{ hours}$$

Let the time spent for the return be t_2:

$$\text{Average rate} = \frac{\text{Total distance traveled}}{\text{Total time}} = \frac{240 \cdot 2}{4 + t_2} = 48$$

$$\frac{480}{4 + t_2} = 48$$
$$480 = 48 \cdot (4 + t_2) \quad \text{Divide by 48.}$$
$$10 = 4 + t_2$$
$$t_2 = 6$$

$$r_2 = \frac{D_2}{t_2} = \frac{240}{6} = 40 \text{ miles per hour}$$

3. **B** $\quad d_1 + d_2 = d_{total}$
$$r_1 \cdot t + r_2 \cdot t = d_{total} \quad \text{Notice that } t \text{ is the same for both.}$$
$$65t + 55t = 300$$
$$120t = 300$$
$$t = 2.5 \text{ hours}$$

4. **B** $t = \dfrac{D}{r}$

The time for part 1: $t = \dfrac{180}{60} = 3$ hours

The time for part 2: $t = \dfrac{120}{30} = 4$ hours

	Rate	×	Time	=	Distance
Part 1	60		3		180
Part 2	30		4		120
Total			7		300

Average rate $= \dfrac{\text{Total distance traveled}}{\text{Total time}}$

Average rate $= \dfrac{300}{7} = 42\dfrac{6}{7} \approx 43$

5. **E** Let the cost to produce in Country *B* be *x*. The cost to produce in Country *A* is 20% higher than *x*, which is $x + 0.2x = 1.2x$.

Average cost per phone $= \dfrac{\text{Total cost}}{\text{Total number of phones}}$

Average cost per phone $= \dfrac{60{,}000 \cdot 1.2x + 48{,}000 \cdot x}{60{,}000 + 48{,}000} = 10$

$$72{,}000x + 48{,}000x = 108{,}000 \cdot 10$$

$$120{,}000x = 1{,}080{,}000$$

$$120x = 1{,}080$$

$x = \$9$ is the cost per phone in Country *B*.

$1.2x = 1.2 \cdot 9 = \$10.80$ is the cost per phone in Country *A*.

6. **C**

	Rate	×	Time	=	Distance
Carl	12		t		$12t$
Jen	8		t		$8t$
Total					15

$$d_1 + d_2 = d_{\text{total}}$$

$$12t + 8t = 15$$

$$20t = 15$$

$$t = \dfrac{15}{20} \text{ hours}$$

$$\dfrac{15}{20} \cdot 60 = 45 \text{ minutes}$$

7. **E** When the hare catches up with the tortoise, they are at the same distance from their starting point.

$d_1 = d_2$

Let the time the tortoise walked be t.

Since the hare started 2 hours later, its time is $(t - 2)$.

$$d_1 = d_2$$
$$r_1 \cdot t = r_2 \cdot (t - 2)$$
$$2t = 6(t - 2)$$
$$2t = 6t - 12$$
$$4t = 12$$
$$t = 3 \text{ hours}$$

8. **B** Rate $= \dfrac{\$c}{h \text{ hours}} = \dfrac{c}{h}$

$\dfrac{\$}{\text{hour}} \cdot \text{hours} = \$$

$\dfrac{c}{h} \cdot k = \dfrac{c \cdot k}{h}$

9. **C**

	Rate	×	Time	=	Distance
Fast	50		t		$50t$
Slow	40		$t + \frac{1}{4}$		$40\left(t + \frac{1}{4}\right)$
Total					

$$d_1 = d_2$$
$$50t = 40 \cdot \left(t + \frac{1}{4}\right)$$
$$50t = 40t + 10$$
$$10t = 10$$
$$t = 1 \text{ hour}$$

If the fast car traveled 1 hour between the cities:

$D = r \cdot t = 50\dfrac{\text{mi}}{\text{hr}}$ 1 hr $= 50$ miles

10. **D** Average rate $= \dfrac{\text{Friends added}}{\text{Time}}$

$r = \dfrac{F}{t} = \dfrac{m}{7 \text{ days}} = \dfrac{m}{7}$ friends per day.

Since we are looking at a period of n days:

Total friends added $= \dfrac{m}{7} \cdot n = \dfrac{mn}{7}$

Daria had k friends to start. So the total number of friends at the end of n days is:

$k + \dfrac{mn}{7}$

11. **E** $d_1 = d_2$

Let the time to go to work be t.

Since she spent $\frac{1}{3}$ hours less on the way back, $\left(t - \frac{1}{3}\right)$ is the time it took her to go back home.

$$d_1 = d_2$$
$$r_1 \cdot t = r_2 \cdot \left(t - \frac{1}{3}\right)$$
$$14t = 18 \cdot \left(t - \frac{1}{3}\right)$$
$$14t = 18t - 6$$
$$6 = 4t$$
$$t = \frac{6}{4}$$
$$t = 1.5 \text{ hours}$$

$D = r \cdot t = 1.5 \cdot 14 = 21$ miles (one way)

Commute $= 21 \cdot 2 = 42$ miles

12. **C** $\qquad d_1 + d_2 = d_{total}$

$\qquad r_1 \cdot t + r_2 \cdot t = d_{total}$ $\qquad$ Notice that t is the same for both.

$$120 \cdot t + 150 \cdot t = 90$$
$$270 \cdot t = 90$$
$$t = \frac{1}{3}$$

They meet after $\frac{1}{3}$ of an hour.

$D = r \cdot t = 120 \cdot \frac{1}{3} = 40$ miles

SECTION 4.3: WORK PROBLEMS

Work problems are essentially rate problems. They can be solved using the same formula.

$$\text{Rate} \cdot \text{Time} = \text{Work done}$$

The rate for work problems is work completed per unit time.

$$\frac{\text{Work}}{\text{Hour}} \cdot \text{Hours} = \text{Work done}$$

Notice that time and rate are inversely proportional. As the rate increases, the amount of time needed to complete the work decreases.

Work Done Collectively

In most work problems, the rate at which a certain job can be completed by certain machines or persons is given and the rate at which the work can be completed in collaboration needs to be computed.

In order to combine the work done by each machine, we calculate the portion (fraction) of work completed per unit time for each machine.

For example, pipe A can fill a pool in 4 hours. Therefore, it can fill $\frac{1}{4}$ of the pool in 1 hour.

For example, pipe B can fill the same pool in 6 hours. Therefore, it can fill $\frac{1}{6}$ of the pool in 1 hour.

We can add the two rates to determine what fraction of the pool can be filled in 1 hour. If the pipes work together, $\frac{1}{4} + \frac{1}{6} = \frac{3}{12} + \frac{2}{12} = \frac{5}{12}$ of the pool can be filled in 1 hour.

If the pipes can fill $\frac{5}{12}$ of the pool in 1 hour, they can fill the entire pool in $\frac{12}{5} = 2\frac{2}{5}$ hours. The proportion below shows you how to calculate the time.

$$\frac{\frac{5}{12}}{1} = \frac{\frac{12}{12}}{x}$$

Note that the proportion will always give you the reciprocal of the sum of the portions of work done.

The fundamental relationship to solve these types of work problems is

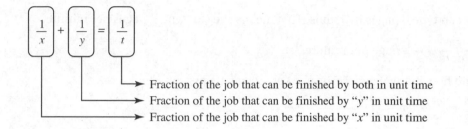

In this relationship, the x is how long the first machine/person needs to finish the job alone. The y is how long the second machine/person needs to finish the job alone. The t is how long both need to finish the job together.

SAMPLE PROBLEMS

EXAMPLE 1

If Juan can read 80 pages in 24 minutes, how long would it take to finish a book that has 360 pages?

Rate $\cdot$ Time = Work done

Juan's rate = $\frac{80}{24} = \frac{10}{3}$ pages/minute

$\frac{10}{3} \cdot t = 360$

$t = 360 \cdot \frac{3}{10} = 36 \cdot 3 = 108$ minutes

Alternatively, since the number of pages is directly proportional to the reading time, we could also set up a proportion:

$\frac{\text{Pages}}{\text{Minute}} \to \frac{80}{24} = \frac{360}{t}$ Cross multiply.

$\qquad 80 \cdot t = 24 \cdot 360$ Divide by 80 before multiplying in order to simplify.

$t = \frac{24 \cdot 360}{80} = \frac{3 \cdot 36}{1} = 108$ minutes

EXAMPLE 2

If Terry can type 5 pages in 7 minutes and Kerry can type 6 pages in 5 minutes, how long would it take them to type 335 pages if they work together?

Terry can type 5 pages in 7 mins. That means he can type $\frac{5}{7}$ of a page in one minute.

$$\left(\frac{5 \text{ pages}}{7 \text{ minute}}\right) \rightarrow \frac{5}{7} \text{ page per minute}$$

Kerry can type 6 pages in 5 mins. That means she can type $\frac{6}{5}$ pages per minute.

$$\left(\frac{6 \text{ pages}}{5 \text{ minute}}\right) \rightarrow \frac{6}{5} \text{ page per minute}$$

To find how many pages Terry and Kerry can type per minute,

$$\frac{5}{7} + \frac{6}{5} = \frac{25}{35} + \frac{42}{35} = \frac{67}{35}$$

Together they can type $\frac{67}{35}$ pages in 1 minute.

At this stage, we can either use the general rate formula or set up a proportion to find the total time to type 335 pages.

Option 1:   Rate · Time = Work

$$\frac{67}{35} \cdot t = 335 \qquad \text{Multiply both sides by } \frac{35}{67}.$$

$$t = 335 \cdot \frac{35}{67} = 175 \text{ minutes}$$

Option 2: $$\frac{\frac{67}{35}}{1} = \frac{335}{t}$$

$$\frac{67}{35} \cdot t = 335 \qquad \text{Multiply both sides by } \frac{35}{67}.$$

$$t = 335 \cdot \frac{35}{67} = 175 \text{ minutes}$$

SECTION 4.3—PRACTICE PROBLEMS

1. Printing press *A* can finish a certain job three times as fast as printing press *B*. When working together, they can finish the job in 12 hours. How many hours will it take press *B* to finish the job when working alone?

   (A) 4
   (B) 16
   (C) 36
   (D) 48
   (E) 64

2. Machine *A* working alone can complete a job in $2\frac{1}{2}$ hours. Machine *B* working alone can do the same job in $3\frac{3}{4}$ hours. How long will it take for both machines working together at their respective constant rates to complete the job?

   (A) 1 hour 30 minutes
   (B) 1 hour 45 minutes
   (C) 2 hours
   (D) 2 hour 15 minutes
   (E) 2 hours 30 minutes

3. Faucet 1 working alone can fill a tub in 20 minutes. Faucet 2 working alone can do the same job in 15 minutes. The hole at the bottom of the tub can empty a full tub in 30 minutes. If both faucets are working at their constant rates given above and the hole is not covered, how long does it take to fill an empty tub?

   (A) 5 minutes
   (B) 6 minutes
   (C) 10 minutes
   (D) 12 minutes
   (E) 20 minutes

4. When working alone, Dominic takes twice as much time as Nick does to mow a lawn. When working together, they can mow half of the same lawn in 15 minutes. How long does it take Dominic to mow the lawn by himself?

   (A) 90 minutes
   (B) 60 minutes
   (C) 45 minutes
   (D) 30 minutes
   (E) 20 minutes

5. Austin takes 5 minutes to prepare and seal one wedding invitation in its envelope. If his fiancée Felicity can do the same job at 3 minutes per invitation, how many envelopes can they prepare in 30 minutes together?

   (A) 8
   (B) 10
   (C) 12
   (D) 14
   (E) 16

6. Bottling Machine A1000 can fill small bottles in 2 seconds or large bottles in 3 seconds. A plant is to be designed to fill a minimum of 20,000 small bottles and 15,000 large bottles in one 8-hour shift. Assuming no downtime, how many A1000s are needed?

   (A) 1
   (B) 2
   (C) 3
   (D) 4
   (E) 5

7. Machine *A* can finish a job in 20 hours. Machine *B* can finish the same job in 45 hours. How long would it take to finish half the job if 2 machine *A*'s and 3 machine *B*'s are used simultaneously?

(A) 6

(B) 5

(C) 4

(D) 3

(E) 2

8. If *S* software engineers working *h* hours a day can finish coding a computer program in *d* days, how many days would it take *R* engineers working *i* hours a day to finish the same job?

(A) $\dfrac{Shd}{Ri}$

(B) $\dfrac{Shi}{Rd}$

(C) $\dfrac{Si}{Rdh}$

(D) $\dfrac{Rd}{Shi}$

(E) $\dfrac{Ri}{Shd}$

SECTION 4.3—SOLUTIONS

1	2	3	4	5	6	7	8
D	A	D	A	E	C	D	A

1. **D** Let the time it takes *A* to finish the job be *t*. Press *B* would finish the job in $(3t)$ hours. Set up the basic relationship for work problems.

$$\frac{1}{A} + \frac{1}{B} = \frac{1}{t} + \frac{1}{3t} = \frac{1}{12}$$

Make the denominators equal by expanding the first fraction by 3

$$\frac{3}{3t} + \frac{1}{3t} = \frac{1}{12} \qquad \text{Add the fractions.}$$

$$\frac{4}{3t} = \frac{1}{12} \qquad \text{Cross multiply.}$$

$$12 \cdot 4 = 3t \qquad \text{Divide both sides by 3.}$$

$$\frac{12 \cdot 4}{3} = t \qquad \text{Simplify.}$$

$$t = 4 \cdot 4 = 16$$

This means press *A* can finish the job in 16 hours. So, press *B* can finish the job in $16 \cdot 3 = 48$ hours.

2. **A** $\dfrac{1}{A} + \dfrac{1}{B} = \dfrac{1}{2\frac{1}{2}} + \dfrac{1}{3\frac{3}{4}} = \dfrac{1}{t}$

$$\frac{1}{\frac{5}{2}} + \frac{1}{\frac{15}{4}} = \frac{1}{t}$$

$$\frac{2}{5} + \frac{4}{15} = \frac{6}{15} + \frac{4}{15} = \frac{10}{15} = \frac{2}{3} = \frac{1}{t}$$

$$t = \frac{3}{2} \text{ hours}$$

$$\frac{3}{2} \text{ hours} = 1 \text{ hour } 30 \text{ minutes}$$

3. **D** $\frac{1}{F_1} + \frac{1}{F_2} - \frac{1}{H} = \frac{1}{20} + \frac{1}{15} - \frac{1}{30} = \frac{1}{t}$

Notice we are subtracting the water let out by the hole. Convert the fractions so they all have a common denominator.

$\frac{3}{60} + \frac{4}{60} - \frac{2}{60} = \frac{5}{60} = \frac{1}{12} = \frac{1}{t}$

$t = \frac{12}{1} = 12$ minutes

4. **A** When working together, the entire lawn takes 30 minutes to mow.

Let the time Nick takes to mow the entire lawn be t.

The time Dominic takes to mow the entire lawn becomes $2t$.

$\frac{1}{D} + \frac{1}{N} - \frac{1}{2t} + \frac{1}{t} = \frac{1}{30}$

$\frac{1}{2t} + \frac{2}{2t} = \frac{1}{30}$

$\frac{3}{2t} = \frac{1}{30}$

$2t = 90$

$t = 45$ minutes

Dominic would take $2 \cdot t = 2 \cdot 45 = 90$ minutes to mow the lawn by himself.

5. **E** Do not use the long formula. Just set up simple proportions in this case because the amount of work is being asked and the total time is given.

Austin: $\frac{5 \text{ min}}{1 \text{ env}} = \frac{30 \text{ min}}{x}$

$x = 6$ envelopes in 30 minutes

Felicity: $\frac{3 \text{ min}}{1 \text{ env}} = \frac{30 \text{ min}}{y}$

$y = 10$ envelopes in 30 minutes

Total envelopes = 16

Alternatively, if you prefer using the formula.

Rate · Time = Work done

Austin: $\frac{1 \text{ env}}{5 \text{ min}} \cdot 30 \text{ min} = 6$ envelopes

Felicity: $\frac{1 \text{ env}}{3 \text{ min}} \cdot 30 \text{ min} = 10$ envelopes

$6 + 10 = 16$ envelopes

6. **C** The total time needed to fill 20,000 small bottles is

Rate × Time = Work done

$\frac{1 \text{ bottle}}{2 \text{ seconds}} \cdot t = 20,000$

$t = 2 \cdot 20,000 = 40,000$ seconds

The total time needed to fill 15,000 large bottles is

$t = 3 \cdot 15,000 = 45,000$ seconds

One A1000 can finish this job in 85,000 seconds.

Converting seconds to hours:

$$85{,}000 \text{ seconds} \cdot \frac{1 \text{ hour}}{3{,}600 \text{ seconds}} = 23.88 \text{ hours}$$

Since the job needs to be done in 8 hours, we need

$$\frac{23.88 \text{ hours}}{8 \text{ hours}} \approx 3 \text{ A1000s}$$

7. **D** 1 machine A can finish $\frac{1}{20}$ of the job in 1 hour.

2 machine A's can finish $\frac{2}{20}$ of the job in 1 hour.

1 machine B can finish $\frac{1}{45}$ of the job in 1 hour.

3 machine B's can finish $\frac{3}{45}$ of the job in 1 hour.

$$\frac{2}{20} + \frac{3}{45} = \frac{1}{10} + \frac{1}{15} = \frac{3}{30} + \frac{2}{30} = \frac{5}{30} = \frac{1}{6}$$

So all machines can finish the job in 6 hours.

The question is asking for the time to finish half of the job, which is 3 hours.

8. **A** Find the total amount of work first.

$$S \text{ engineers} \cdot h \frac{\text{hours}}{\text{day}} \cdot d \text{ days} = Shd \text{ hours of work}$$

The new number of engineers is R and the workday is i hours long. Let the new number of days be x. The total amount of work stays the same.

$$Shd = Rix$$
$$x = \frac{Shd}{Ri}$$

SECTION 4.4: MIXTURE PROBLEMS

In mixture problems, usually two or more products with different characteristics are mixed. Certain characteristics of the resultant mixture are asked about. In some problems, the composition will be changed via adding or removing one of the ingredients.

The fundamental relationship to solve mixture problems is

> % of product · Amount of mixture = Amount of product in that mixture

Once you mix, you get the following equation:

> Product in mixture 1 + Product in mixture 2 = Total product in the resulting mix

Example: How many liters of a 25% salt solution need to be mixed with 15 liters of 10% salt solution to create a 16% salt solution?

<table>
<tr><td></td><td>% of salt</td><td>× mixture</td><td>= Amount of salt</td></tr>
<tr><td>Solution 1</td><td>25%</td><td>m</td><td>$0.25m$</td></tr>
<tr><td>Solution 2</td><td>10%</td><td>15</td><td>1.5</td></tr>
<tr><td>Mixture</td><td>16%</td><td>$m + 15$</td><td>$0.16(m + 15)$</td></tr>
</table>

NOTICE

Mixture problems are mathematically very similar to the rate problems we reviewed in Section 4.2, where $d_1 + d_2 = d_t$.

Salt in solution 1 + Salt in solution 2 = Total salt

$0.25m + 1.5 = 0.16(m + 15)$	Multiply each term by 100 to avoid decimals.
$25m + 150 = 16(m + 15)$	Distribute 16.
$25m + 150 = 16m + 240$	Rearrange the terms.
$25m - 16m = 240 - 150$	
$9m = 90$	Divide by 9.
$m = 10$ liters	

Mixture Problems Involving Cost

If the mixture problem is about cost, then use the following equation:

Price × Amount = Cost

Once you mix, you get:

$Cost_1 + Cost_2$ = Total cost

SAMPLE PROBLEMS

EXAMPLE 1

The cost of cashews is $8.50 per pound, and the cost of peanuts is $5 per pound. If a grocer wants to produce a 10-pound mixture that costs $6.40 per pound, how many pounds of cashews should she use in the mixture?

	Price	× Amount	=	Cost
Cashews	$8.50/lb	a		$8.5a$
Peanuts	$5.00/lb	$(10 - a)$		$5(10 - a)$
Mixture	$6.40/lb	10		$64

$Cost_1 + Cost_2$ = Total cost	
$8.5a + 5(10 - a) = 64$	Distribute 5 over the parentheses.
$8.5a + 50 - 5a = 64$	Combine like terms, and subtract 50 from each side.
$3.5a = 14$	Divide by 3.5.
$a = 4$ pounds of cashews	

EXAMPLE 2

A piggy bank contains 50 coins consisting of quarters and dimes. Their total value is $7.40. How many quarters are there?

	Price	× Amount	=	Cost
Quarters	25¢/piece	x		$25x$
Dimes	10¢/piece	$(50 - x)$		$10(50 - x)$
Total		50		740¢

$$\text{Cost}_1 + \text{Cost}_2 = \text{Total cost}$$

Write the total cost equation in cents to avoid decimals.

$$25x + 10(50 - x) = 740$$

Distribute 10 over the parentheses.

$$25x + 500 - 10x = 740$$

Combine like terms, and subtract 500 from each side.

$$15x = 240$$

Divide by 15.

$$x = 16 \text{ quarters}$$

SECTION 4.4—PRACTICE PROBLEMS

1. Giant brand sports drink contains 15% fruit juice by volume. How many liters of pure fruit juice need to be added to 20 liters of the sports drink to increase the juice percentage to 20%?

   (A) 1 liter

   (B) $1\frac{1}{4}$ liters

   (C) $1\frac{1}{2}$ liters

   (D) $1\frac{3}{4}$ liters

   (E) 2 liters

2. 25 liters of 10% salt water are mixed with 15 liters of 20% salt water. The 10 liters of water are added to the mix. What is the resulting percentage of salt?

   (A) 10.50%

   (B) 11.00%

   (C) 13.75%

   (D) 15.00%

   (E) 15.25%

3. A drink is 12% alcohol by volume. How much water do you need to add to a 300 cc drink to decrease the alcohol content to 8% by volume?

   (A) 50 cc

   (B) 75 cc

   (C) 100 cc

   (D) 125 cc

   (E) 150 cc

4. How much water needs to be evaporated from a 30-gallon 20% salt solution to result in a mixture of 40% salt?

   (A) 5

   (B) 10

   (C) 12

   (D) 15

   (E) 17

5. When 30 pounds of sugar are added to a mixture of 40% sugar, the concentration increases to 50% sugar. What is the weight of the original mixture?

(A) 100 pounds
(B) 125 pounds
(C) 150 pounds
(D) 175 pounds
(E) 200 pounds

6. A mixture of red beans and black beans are to be prepared. The price of red beans is $2 per pound, and the price of black beans is $3 per pound. What is the ratio of red beans to black beans if the mixture is to be sold for $2.75 per pound?

(A) $1:1$
(B) $1:2$
(C) $1:3$
(D) $2:3$
(E) $3:4$

7. Equal amounts of water and acid are added to 40 quarts of 80% acid solution. The new mixture has a concentration of 70%. How much acid is in the resulting solution?

(A) 10 quartz
(B) 18 quartz
(C) 32 quartz
(D) 40 quartz
(E) 42 quartz

8. Coffee beans from Ethiopia cost a dollars per pound, and beans from Guatemala cost b dollars per pound. In a certain mixture, there are three times as many beans from Ethiopia as beans from Guatemala. What is the cost per pound of this mixture?

(A) $\dfrac{b + 3a}{4}$
(B) $\dfrac{a + 3b}{4}$
(C) $\dfrac{a + b}{2}$
(D) $\dfrac{3a + b}{2}$
(E) $a + 3b$

1	2	3	4	5	6	7	8
B	B	E	D	C	C	E	A

1. **B** The amount of juice in the original drink is

   $0.15 \cdot 20$ liters = 3 liters

   Let the amount of juice to be added be x.

   Notice that the % juice of pure juice is 100%.

	% of juice $\times$	Mixture $=$	Total juice
Drink 1	15%	20	3
Juice	100%	x	x
Mixture	20%	$20 + x$	$0.2(20 + x)$

   $0.2(20 + x) = 3 + x$

   $\quad 4 + 0.2x = 3 + x$

   $\qquad\quad 1 = 0.8x$

   $\qquad\quad x = 1.25$ liters

   Alternatively, you can set up a proportion.

   $$\text{Concentration} = \frac{\text{Amount of juice}}{\text{Total mixture}}$$

   Since the new drink is 20% juice, $\frac{20}{100} = \frac{1}{5}$.

   $\dfrac{1}{5} = \dfrac{3 + x}{20 + x}$ Remember to add the juice to the total mixture in the denominator.

   $20 + x = 15 + 5x$

   $\quad\; 4x = 5$

   $\qquad x = \dfrac{5}{4} = 1.25$ liters

2. **B** The amount of salt in the 1st mix is $0.10 \cdot 25 = 2.5$.

   The amount of salt in the 2nd mix is $0.20 \cdot 15 = 3.0$.

   The amount of water to be added is 10 liters.

   Notice that the percent salt for pure water is 0%.

	% of salt $\times$	Mixture $=$	Total salt
Mix 1	10%	25	2.5
Mix 2	20%	15	3.0
Water	0%	10	0
Mixture		50	5.5

   $$\text{Concentration} = \frac{\text{Amount of salt}}{\text{Total mixture}}$$

   $$\frac{2.5 + 3.0}{25 + 15 + 10} = \frac{5.5}{50} = 0.11 = 11\%$$

3. **E** The amount of alcohol in the original drink is $0.12 \cdot 300 = 36$ cc

Let the amount of water to be added be x.

Notice that the % alcohol of pure water is 0%.

	% of alcohol	×	Mixture	=	Total alcohol
Drink 1	12%		300		36
Water	0%		x		0
Mixture	8%		$300 + x$		$0.08(300 + x)$

$$0.08(300 + x) = 36$$
$$24 + 0.08x = 36$$
$$0.08x = 12$$
$$x = \frac{1200}{8} = 150 \text{ cc}$$

Alternatively, you can set up a proportion.

$$\text{Concentration} = \frac{\text{Amount of alcohol}}{\text{Total mixture}}$$

Since the new drink is 8% alcohol, $\frac{8}{100} = \frac{2}{25}$.

$$\frac{2}{25} = \frac{36}{300 + x}$$
$$2(300 + x) = 25 \cdot 36$$
$$600 + 2x = 900$$
$$2x = 300$$
$$x = 150 \text{ cc}$$

4. **D** The amount of salt in the 1st mix is $0.20 \cdot 30 = 6$.

Since the amount of salt will not change, the resulting solution will still have 6 gallons of salt and less water.

Let the water to be evaporated be x.

$$\text{Concentration} = \frac{\text{Amount of salt}}{\text{Total mixture}}$$

$$\frac{6}{30 - x} = \frac{40}{100}$$
$$\frac{6}{30 - x} = \frac{2}{5}$$
$$6 \cdot 5 = 2(30 - x)$$
$$30 = 60 - 2x$$
$$2x = 30$$
$$x = 15 \text{ gallons}$$

5. **C** Let the weight of mixture 1 be x.

The amount of sugar in the original mix is $0.4 \cdot x = 0.4x$

	% of sugar ×	Mixture =	Total sugar
Mixture 1	40%	x	$0.4x$
Sugar	100%	30	30
Final Mixture	50%	$300 + x$	$0.5(30 + x)$

Sugar 1 + Sugar 2 = Total sugar

$$0.4x + 30 = 0.5(30 + x)$$
$$0.4x + 30 = 15 + 0.5x$$
$$15 = 0.1x$$
$$x = 150 \text{ pounds}$$

6. **C** In this problem, we will need to use two different variables since the amount of red beans and black beans are not defined in terms of each other. In fact, that is the question.

Let the amount of red beans in the mixture be x.
The cost of red becomes $2 times x.
Let the amount of black beans in the mixture be y.
The cost of black becomes $3 times y.
Question: $\frac{x}{y} = ?$

	Price ×	Amount =	Cost
Red	$2/lb	x	$2x$
Black	$3/lb	y	$3y$
Mixture	$2.75/lb	$(x + y)$	$2.75(x + y)$

$cost_1 + cost_2 = $ total cost

$$2x + 3y = 2.75 \cdot (x + y)$$
$$2x + 3y = 2.75x + 2.75y$$
$$3y - 2.75y = 2.75x - 2x$$
$$0.25y = 0.75x$$
$$\frac{x}{y} = \frac{0.25}{0.75} = \frac{1}{3}$$

7. **E** The amount of acid in the original mix is $40 \cdot 0.8 = 32$ quarts.

The amount of water in the original mix is $40 - 32 = 8$ quarts.

The new concentration is $70\% = \frac{70}{100} = \frac{7}{10}$.

$$\text{Concentration} = \frac{\text{Amount of acid}}{\text{Total mixture}}$$

$$\frac{32 + x}{40 + x + x} = \frac{7}{10}$$
$$10(32 + x) = 7(40 + 2x)$$
$$320 + 10x = 280 + 14x$$
$$40 = 4x$$
$$x = 10 \text{ quarts} \qquad \text{Go back to the question and find out what it is asking.}$$

The resulting solution has $10 + 32 = 42$ quarts of acid.

8. **A** Let the amount of beans from Guatemala in the mixture be x. So the total cost of beans from Guatemala becomes x times b \$/lb.

Let the amount of beans from Ethiopia in the mixture be $3x$. So the total cost of beans from Ethiopia becomes $3x$ times a \$/lb.

$$\text{Average cost} = \frac{\text{Total cost}}{\text{Total mixture}}$$

$$= \frac{x \cdot b + 3x \cdot a}{x + 3x}$$

$\dfrac{x \cdot b + 3x \cdot a}{x + 3x}$ Factor out an x from the numerator and the denominator.

$\dfrac{x \cdot (b + 3 \cdot a)}{4x}$ Simplify the x's

$\dfrac{(b + 3a)}{4}$

The cost per pound of the mixture is $\dfrac{(b + 3a)}{4}$.

SECTION 4.5: INVESTMENT/INTEREST PROBLEMS

The amount of **simple interest** can be calculated as

$$\text{Interest} = \text{Principal} \cdot \text{Yearly interest rate} \cdot \text{Time in years}.$$

This is commonly written as

$$I = P \cdot r \cdot t$$

If interest is **compounded**, the resulting balance (principal + interest) can be found as follows:

$$\text{Balance} = \text{Principal} \cdot \left(1 + \frac{\text{Yearly interest}}{\text{Number of periods per year}}\right)^{\text{Number of periods per year} \times \text{time in years}}$$

The formula is

$$B = P \cdot \left(1 + \frac{r}{n}\right)^{n \cdot t}$$

SAMPLE PROBLEMS

> **EXAMPLE 1**
>
> How much interest is earned if \$12,000 is invested for 9 months at 4 percent simple annual interest?

$I = P \cdot r \cdot t$

$I = \$12{,}0.500 \cdot 0.04 \cdot \dfrac{9}{12}$ Notice that we convert 9 months into $\dfrac{9}{12} = \dfrac{3}{4}$ years.

$I = \$360$

EXAMPLE 2

What expression could be used to calculate the balance after 4 years if $2,000 is invested at 6 percent interest compounded semiannually?

$$B = P \cdot \left(1 + \frac{r}{n}\right)^{n \cdot t}$$

$$B = \$2,000 \cdot \left(1 + \frac{0.06}{2}\right)^{2 \cdot 4}$$ Since the compounding is done semiannually, $n = 2$.

$$B = \$2,000 \cdot (1 + 0.03)^8 = \$2,000 \cdot (1.03)^8$$

Partial Investments (Portfolio Returns)

If a certain amount of money is divided into portions to invest at different interest rates:

$$\text{Interest}_1 + \text{Interest}_2 = \text{Total interest}$$

SAMPLE PROBLEM

Eve invested $3,000 at a certain simple interest rate and $4,000 at a simple interest rate that is 2% higher. What was the interest in her first investment if the total interest she earned was $360?

$\text{Interest}_1 + \text{Interest}_2 = \text{Total interest}$

	Principal	×	Interest rate	=	Interest ($)
Inv. 1	$3,000		r		$3,000r$
Inv. 2	$4,000		$(r + 0.02)$		$4,000(r + 0.02)$
Total					360

$\$3,000r + \$4,000(r + 0.02) = \$360$

$3,000r + 4,000r + 80 = 360$

$7,000r = 280$

$r = \dfrac{280}{7,000} = \dfrac{280 \div 70}{7,000 \div 70} = \dfrac{4}{100} = 4\%$

SECTION 4.5—PRACTICE PROBLEMS

1. 80% of $10,000 was invested in stock A, which lost 20% of its value. The rest was invested in stock B, which gained 30% in value. What is the overall percent loss or gain from this investment?

   (A) 20% gain
   (B) 10% gain
   (C) No loss or gain
   (D) 10% loss
   (E) 20% loss

2. A portfolio contains 60% bonds yielding 4% per year and 40% stocks yielding 8% per year. What is the expected yield of the portfolio?

(A) 4.6%
(B) 4.8%
(C) 5.2%
(D) 5.6%
(E) 6.0%

3. Shelly deposits $12,000 in a bank that pays 4% simple interest per year. If Shelly's interest earnings are taxed at 15%, how much total money will she have at the end of one year?

(A) $12,480
(B) $12,408
(C) $12,096
(D) $408
(E) $384

4. Nader has a credit card with an annual interest rate of 24% and a balance of $3,200. How much interest will he owe in one month?

(A) $3,264
(B) $640
(C) $64
(D) $32
(E) $24

5. Greg invests $\frac{1}{3}$ of his $15,000 at 4% simple interest rate. He invests $\frac{3}{4}$ of the remaining amount at 6% simple interest rate. Finally, he keeps the remaining amount in cash with no return. What is his average rate of return at the end of the year?

(A) $4\frac{1}{3}\%$
(B) $4\frac{1}{2}\%$
(C) 5%
(D) $5\frac{1}{2}\%$
(E) $5\frac{1}{3}\%$

6. If Teresa invests T dollars at an annual simple interest rate of $i\%$, how much money she will have at the end of z years?

(A) $T \cdot i \cdot z$
(B) $\dfrac{T \cdot i \cdot z}{100}$
(C) $T \cdot \dfrac{i}{z}$
(D) $\dfrac{T + Tiz}{100}$
(E) $\dfrac{100T + Tiz}{100}$

7. Sue has twice as much money as Carrie. If Carrie makes 5.5% simple interest per year and Sue makes 5% simple interest per year, what is the ratio of the interest earned by Carrie to the interest earned by Sue in one year?

(A) $\frac{11}{10}$

(B) $\frac{11}{20}$

(C) $\frac{10}{11}$

(D) $\frac{20}{11}$

(E) $\frac{1}{2}$

8. Global Earnings Inc. operates in two countries. Country A's tax rate is three times as much as country B's tax rate. If Global's earnings in country B are twice as much as its earnings in country A, what is the ratio of Global's average tax rate to country A's tax rate?

(A) $\frac{3}{5}$

(B) $\frac{5}{6}$

(C) $\frac{1}{3}$

(D) $\frac{2}{3}$

(E) $\frac{5}{9}$

SECTION 4.5—SOLUTIONS

1	2	3	4	5	6	7	8
D	D	B	C	A	E	B	E

1. **D** Gain/loss from a stock investment is calculated exactly the same way as we calculate interest. The only difference is that the loss is shown as a negative value.

80% of $10,000 = 0.8 · $10,000 = $8,000

So $8,000 is invested in stock A and $2,000 is invested in stock B.

Gain/loss$_1$ + Gain/loss$_2$ = Total gain or loss

$2,000 \cdot 0.3 - 8,000 \cdot 0.2 = 600 - 1,600 = -1,000$

Therefore the percent loss is $\frac{-1,000}{10,000} \cdot 100 = -10\%$

2. **D** Interest$_1$ + Interest$_2$ = Total interest

$0.6 \cdot 0.04 + 0.4 \cdot 0.08 = 0.056 = 5.6\%$

Alternatively, assume the portfolio starts with $100. That makes bonds $100 \cdot 60\% = \$60$ and stocks $100 - 60 = \$40$.

The total yield is
$60 \cdot 0.04 + 40 \cdot 0.08 = \$5.60$, which is 5.6%.

3. **B** $I = P \cdot r \cdot t$

   $I = 12,000 \cdot 0.04 \cdot 1 = \$480$

   She pays 15% tax on \$480 only.

   Tax $= 0.15 \cdot 480 = \$72$

   Net interest $= \$480 = \$72 = \$408$

   Total net amount $= \$12,000 + \$408 = \$12,408$

4. **C** $I = P \cdot r \cdot t$

   Note the time is 1 month, which is $\frac{1}{12}$ years.

   $I = 3,200 \cdot 0.24 \cdot \frac{1}{12}$ Simplify first.

   $I = 3,200 \cdot 0.02 = \$64$

   Nader will owe \$64 of interest in one month.

5. **A** Average rate of return $= \dfrac{\text{Total Interest}}{\text{Total Investment}}$

   $\dfrac{(P_1 r_1 t) + (P_2 r_2 t) + (P_3 r_3 t)}{15,000}$

   $\$15,000 \times \frac{1}{3} = \$5,000$

   \$5,000 at 4% for 1 year

   Remaining amount $= \$10,000$

   $\frac{3}{4}$ of the remaining $= \$10,000 \times \frac{3}{4} = \$7,500$

   \$7,500 at 6% for 1 year

   \$2,500 at 0% since it is cash with no return

   $\dfrac{(5,000 \cdot 0.04 \cdot 1) + (7,500 \cdot 0.06 \cdot 1) + (0)}{15,000}$

   $\dfrac{200 + 450}{15,000} = 0.04\overline{3} = 4\frac{1}{3}\%$

6. **E** $I = P \cdot r \cdot t$

   $I = T \cdot \dfrac{i}{100} \cdot z$ Notice that $i\%$ translates into $\dfrac{i}{100}$.

   $I = \dfrac{T \cdot i \cdot z}{100}$

   Total amount $= T + \dfrac{T \cdot i \cdot z}{100} = \dfrac{100T + Tiz}{100}$

7. **B** Let Carrie's money be x. Then Sue's money becomes $2x$.

   $\dfrac{\text{Interest earned by Carrie}}{\text{Interest earned by Sue}} = \dfrac{x \cdot 0.055 \cdot 1}{2x \cdot 0.05 \cdot 1}$

   $\dfrac{0.055}{0.10} = \dfrac{55}{100} = \dfrac{11}{20}$

8. **E** Let country B's tax rate be t and country B's earnings be $2E$.

   Country A's tax rate becomes $3t$, and its earnings become E.

   Average tax $= \dfrac{\text{Total tax}}{\text{Total earnings}} = \dfrac{\text{Tax}_A + \text{Tax}_B}{\text{Total earnings}}$

   $\dfrac{t \cdot 2E + 3t \cdot E}{2E + E} = \dfrac{5tE}{3E} = \dfrac{5t}{3}$

   $\dfrac{\text{Average tax rate}}{\text{Country } A\text{'s tax rate}} = \dfrac{\frac{5t}{3}}{3t} = \dfrac{5t}{3} \cdot \dfrac{1}{3t} = \dfrac{5}{9}$

SECTION 4.6: SET PROBLEMS AND VENN DIAGRAMS

A set is a collection of things. These things are called elements or members of the set. The easiest way to solve set problems is to represent the groups by Venn diagrams as shown below.

The total number of elements in set A is $n(A) = x$.
The total number of elements in set B is $n(B) = y$.
The overlap $= b$.

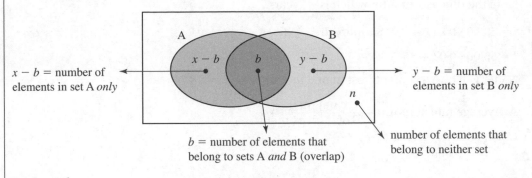

$x - b =$ number of elements in set A *only*

$y - b =$ number of elements in set B *only*

$b =$ number of elements that belong to sets A *and* B (overlap)

number of elements that belong to neither set

In general:

$$\text{Total} = \text{Group1} + \text{Group2} - \text{Both} + \text{Neither}$$

If only the number of elements in either A or B is asked:

$$n(A \text{ or } B) = n(A) + n(B) - n(A \text{ and } B)$$

SAMPLE PROBLEMS

EXAMPLE 1

In a group of 55 people, each person speaks French, German, both, or neither. 17 people speak French, 22 people speak German, and 6 people speak both languages. How many people in this group speak neither German nor French?

Always start by calculating the overlap and placing it in the middle.

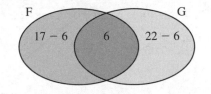

Total $= n(F) + n(G) - n(F \text{ and } G) + \text{Neither} = 17 + 22 - 6 + n = 55$.

Solve for n.

$33 + n = 55$

$\qquad n = 55 - 33 = 22$

22 people speak neither French nor German.

Note that there are $17 - 6 = 11$ French-only speakers and $22 - 6 = 16$ German-only speakers.

EXAMPLE 2

Each of 35 people has an MBA degree, law degree, or both. If 18 people have MBA degrees and 23 people have law degrees, how many people have MBA degrees only?

Let m be the number of people holding both degrees (overlap).

$n(M \text{ or } L) = n(M) + n(L) - n(M \text{ and } L)$

$18 + 23 - m = 35$

$\quad 41 - m = 35$

$m = 6$ is the number of people with both degrees.

So there are $18 - 6 = 12$ people who have MBA degrees only.

Alternatively, use a Venn diagram.

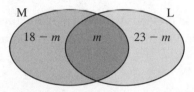

Since the total number of people in the group is 35,

$18 - m + m + 23 - m = 35$

$\qquad\qquad 41 - m = 35$

$m = 6$ is the number of people with both degrees.

So there are $18 - 6 = 12$ people have MBA degrees only.

EXAMPLE 3

A school library contains 1,500 books comprised entirely of fiction and nonfiction, hardcover and softcover. In the library, $\frac{3}{5}$ of the books are hardcover and 3,000 of the books are nonfiction. If 300 of the hardcover books are fiction, how many of the softcover books are nonfiction?

Notice that two separate properties are assigned to each book: fiction/nonfiction and softcover/hardcover. We can organize these data into a table as shown below.

$\frac{3}{5}$ of the books are hardcover.

$\frac{3}{5} \cdot 4,500 = 2,700$ hardcover books

	Softcover	Hardcover	Total
Fiction		300	
Nonfiction	600	2,400	3,000
Total		2,700	4,500

After inputting all given data (shown in bold), we can easily calculate the missing numbers:

Hardcover nonfiction: $2,700 - 300 = 2,400$

Softcover nonfiction: $3,000 - 2,400 = 600$

1. In a certain high school, all students take either American or world history classes. 96 students take American history, and 144 take world history. If 24 students take both classes, what is the ratio of students taking only American history to those taking only world history?

   (A) $\frac{2}{5}$

   (B) $\frac{3}{5}$

   (C) $\frac{2}{3}$

   (D) $\frac{3}{4}$

   (E) $\frac{4}{5}$

2. Out of 90 CEOs surveyed, 32 previously worked in marketing, 40 of them worked in finance, and 18 of them worked in both. How many of them worked in neither marketing nor finance?

   (A) 10

   (B) 18

   (C) 26

   (D) 32

   (E) 36

3. In a group of 69 people having breakfast at a diner, each person ordered coffee, orange juice, or both. The number of people who ordered only coffee is 4 times as many as the number of people who ordered both. If 4 people ordered orange juice only, how many people ordered only coffee?

   (A) 65

   (B) 52

   (C) 26

   (D) 16

   (E) 13

4. A car lot contains 200 vehicles comprised entirely of cars and trucks, used and new. 60% of the vehicles are trucks, and 140 of the vehicles are new. If $\frac{5}{8}$ of the cars are new, how many of the trucks are used?

   (A) 30

   (B) 40

   (C) 50

   (D) 60

   (E) 80

5. 55% of the incoming MBA class play a musical instrument, and 45% of them speak 2 or more languages. 10% play an instrument and speak 2 or more languages. If 20 students neither play an instrument nor speak 2 or more languages, how many of them speak 2 or more languages but do not play an instrument?

(A) 35
(B) 45
(C) 55
(D) 70
(E) 90

6. A survey is conducted among 560 members of a book club about two new books. The number of people who liked both books was equal to the number of people who liked neither. The number of people who liked book *A* is 5 times as many as the number people who liked both. The number of people who liked book *B* is 3 times as many as the number people who liked both. How many people liked book *B* only?

(A) 70
(B) 140
(C) 210
(D) 280
(E) 350

7.

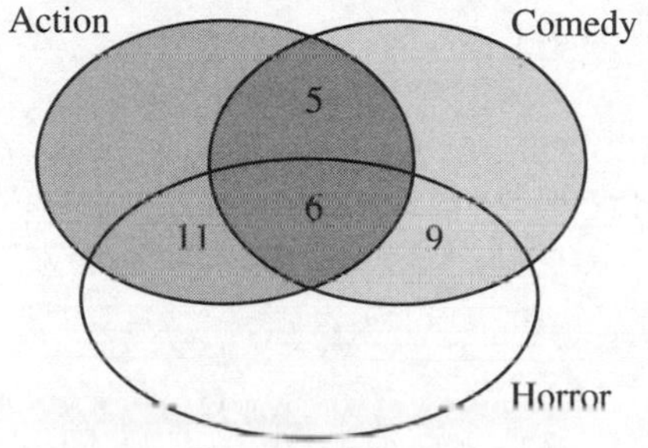

A marketing survey among moviegoers is shown above. If 85 people liked comedy, 42 people liked horror, and 60 people liked action, how many people liked only one type of movie?

(A) 187
(B) 119
(C) 64
(D) 38
(E) 16

8. There are 1,256 new cars in a parking lot. Each car has one or more of the following items: a navigation system, a CD player, or a sunroof. If there are a total of 400 sunroofs and 554 navigation systems and if 1,002 cars have at most 2 items, how many cars have all three items?

(A) 254
(B) 302
(C) 954
(D) 1,208
(E) It cannot be determined from the information given

1	2	3	4	5	6	7	8
B	**E**	**B**	**A**	**D**	**B**	**B**	**A**

1. **B** American history only: $96 - 24 = 72$

   World history only: $144 - 24 = 120$

   $$\frac{72}{120} = \frac{3}{5}$$

   Remember to check the answer choices first. Simplifying the fraction is sufficient in this case.

2. **E** The overlap is already given, so put that in the middle of the Venn diagram.

   If the total number of CEOs with marketing background is 32, then the CEOs with a marketing background only is $32 - 18 = 14$.

   If the total number of CEOs with finance background is 40, then the CEOs with a finance background only is $40 - 18 = 22$.

   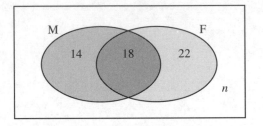

   Neither $= 90 - 14 - 18 - 22 = 36$

   Alternatively, use the formula

   Total $= \text{Group}_1 + \text{Group}_2 - \text{Both} + \text{Neither}$
   $90 = 40 + 32 - 18 + \text{Neither}$
   Neither $= 90 - 40 - 32 + 18 = 36$

3. **B** Let the number of people who ordered both beverages be m. The number of people who ordered coffee only becomes $4m$.

   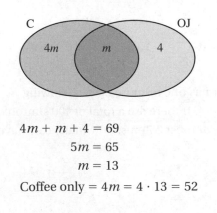

   $$4m + m + 4 = 69$$
   $$5m = 65$$
   $$m = 13$$

   Coffee only $= 4m = 4 \cdot 13 = 52$

4. **A**

	New	Used	Total
Cars	50	30	80
Truck	90	30	120
Total	140	60	200

60% of the vehicles are trucks means $0.6 \cdot 200 = 120$ trucks.

The number of cars is $200 - 120 = 80$.

After finding the total number of cars, $\frac{5}{8}$ of the cars are new, so $\frac{5}{8} \cdot 80 = 50$ new cars.

Since there are 140 new vehicles, $140 - 50 = 90$ new trucks.

So there are $120 - 90 = 30$ used trucks.

5. **D** Instrument only → $55\% - 10\% = 45\%$

Language only → $45\% - 10\% = 35\%$

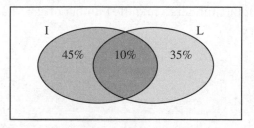

Neither $= 100\% - 45\% - 10\% - 35\% = 10\%$

If 10% of the class is 20 students we can set up a quick proportion to find the total class:

$\frac{10}{100} = \frac{20}{x}$

$x = 200$

Therefore, the number of people who speak 2 or more languages but do not play an instrument is

$35\% \cdot 200 = 70$

6. **B** Let the number of people who liked both be P.

The number of people who liked neither is also P.

The number of people who liked A becomes $5P$.

The number of people who liked A only is $5P - P$.

The number of people who liked B is $3P$.

The number of people who liked B only $3P - P$.

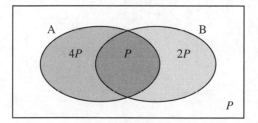

$4P + P + 2P + P = 560$

$8P - 560$

$P = 70$

The number of people who liked B only is $2P = 140$.

Alternatively, you could use the formula without the diagram:

Total = Group 1 + Group 2 − Both + Neither
$560 = 5P + 3P − P + P$
$560 = 8P$
$140 = 2P$

7. **B** Comedy only is $85 − 5 − 6 − 9 = 65$

   Horror only is $42 − 6 − 9 − 11 = 16$

   Action only is $60 − 5 − 6 − 11 = 38$

   $65 + 16 + 38 = 119$ people like one genre only

8. **A** The phrase "at most 2" means the sum of none, only one, or two of the items.

   In other words, "at most" means "not all three."

   So we can simply subtract 1,002 from the total number to find the number of cars that have all three

   $1,256 − 1,002 = 254$

Geometry

5

→ **5.1 LINES, ANGLES, AND TRIANGLES**
→ **5.2 QUADRILATERALS, CIRCLES, AND SOLIDS**
→ **5.3 COORDINATE GEOMETRY AND GRAPHS**
→ **5.4 DATA INTERPRETATION**

SECTION 5.1: LINES, ANGLES, AND TRIANGLES
Lines

- A line is made up of an infinite number of points. It is straight and extends infinitely in both directions.
- Lines can be named individually, such as line ℓ_1, or they can be referred to using two points contained in the line, such as line *KT*.

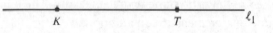

Angles

- In the figure below, line ℓ_1 intersects line ℓ_2 at point *K*. So $\angle a$ and $\angle c$ are vertical angles, and $\angle b$ and $\angle d$ are vertical angles. The measure of angle a ($m\angle a$) equals the measure of angle c. The measure of angle b equals the measure of angle d.

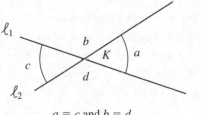

$a = c$ and $b = d$

- The sum of the measures of all angles on one side of a straight line is 180°.

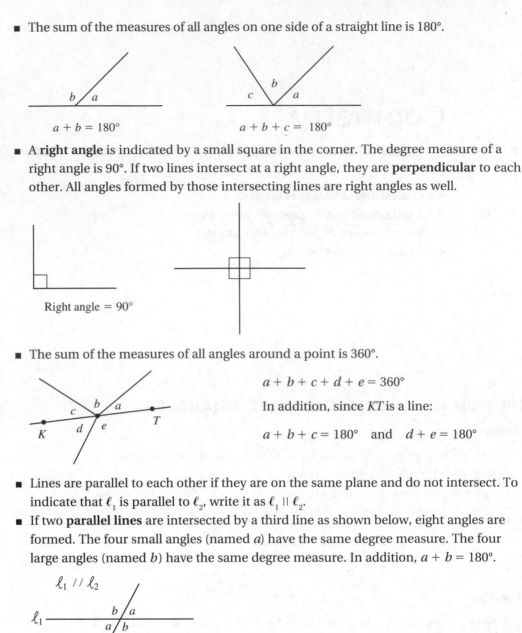

$a + b = 180°$ $\qquad$ $a + b + c = 180°$

- A **right angle** is indicated by a small square in the corner. The degree measure of a right angle is 90°. If two lines intersect at a right angle, they are **perpendicular** to each other. All angles formed by those intersecting lines are right angles as well.

Right angle = 90°

- The sum of the measures of all angles around a point is 360°.

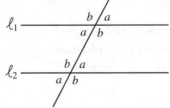

$a + b + c + d + e = 360°$

In addition, since KT is a line:

$a + b + c = 180°$ and $d + e = 180°$

- Lines are parallel to each other if they are on the same plane and do not intersect. To indicate that ℓ_1 is parallel to ℓ_2, write it as $\ell_1 \parallel \ell_2$.

- If two **parallel lines** are intersected by a third line as shown below, eight angles are formed. The four small angles (named a) have the same degree measure. The four large angles (named b) have the same degree measure. In addition, $a + b = 180°$.

$\ell_1 \; / / \; \ell_2$

- A **polygon** is a closed planar figure made up of three or more line segments. Each line segment is called a **side**, and their intersection point is called a **vertex** of the polygon.
- A **diagonal** is a line segment that connects two nonadjacent vertices of a polygon.

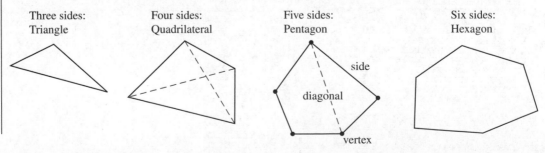

Three sides:
Triangle

Four sides:
Quadrilateral

Five sides:
Pentagon

Six sides:
Hexagon

side

diagonal

vertex

- The **perimeter** of any polygon is the sum of all of its side lengths.
- A **regular polygon** is a polygon that has all equal sides and equal angles:

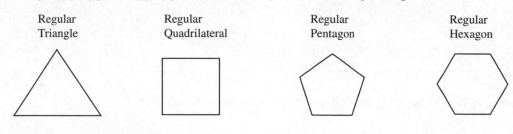

| Regular Triangle | Regular Quadrilateral | Regular Pentagon | Regular Hexagon |

INTERIOR AND EXTERIOR ANGLES

- Interior angles are angles inside a polygon.
- The sum of the interior angles of a triangle is 180°.

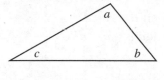

$$a + b + c = 180°$$

- The sum of the interior angles of a quadrilateral is 360°.

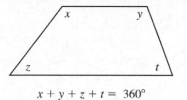

$$x + y + z + t = 360°$$

- In general, the **sum of all interior angles** of any polygon can be found with the formula: $(n - 2) \cdot 180°$, where n is the number of sides.

   Example: For a hexagon, the sum of the interior angles $(6 - 2) \cdot 180° = 4 \cdot 180° = 720°$.

- You can pick one of the vertices of the polygon and draw all diagonals from that point only. The sum of the interior angles of that polygon is equal to 180° times the number of triangles formed.

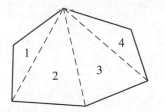

The sum of the interior angles of this hexagon is $4 \cdot 180° = 720°$.

- Exterior angles are angles formed by one side of a polygon and extension of another side. In the diagram below, *a*, *b*, and *c* are exterior angles.

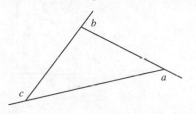

- In a triangle, an exterior angle is the sum of the two remote interior angles.

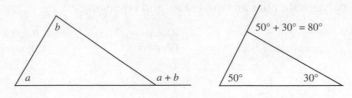

- The sum of all exterior angles of any polygon is always 360°.

Triangles

- In every triangle, the longest side is opposite the largest angle and the shortest side is opposite the smallest angle.

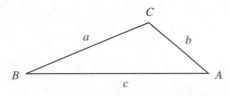

If $b < a < c$, then $B < A < C.$

If $B < A < C$, then $b < a < c.$

- In every triangle, each side is shorter than the sum of the other two sides and longer than the absolute value of their difference.

$$a + b > c > |a - b|$$

RIGHT TRIANGLES

- A triangle with a 90° angle is called a **right triangle**. The longest side is called the **hypotenuse**. The other sides (perpendicular sides) are called the **legs**.
- The **Pythagorean theorem** states that the square of the hypotenuse is equal to the sum of the squares of the legs:

$(\text{hypotenuse})^2 = \text{leg}_1{}^2 + \text{leg}_2{}^2$ or $h^2 = a^2 + b^2$

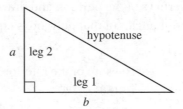

- Several right triangles are frequently seen.

3-4-5 triangle: A triangle with sides proportional to 3 : 4 : 5 is a right triangle.

Examples: The Pythagorean theorem for a triangle with sides 3, 4, 5 is $5^2 = 4^2 + 3^2$. For a triangle with sides, 6, 8, 10, the Pythagorean theorem is $10^2 = 8^2 + 6^2$

5-12-13 triangle: A triangle with sides proportional to $5 : 12 : 13$ is a right triangle.

Examples: The Pythagorean theorem for a triangle with sides 5, 12, 13 is $13^2 = 12^2 + 5^2$.
For a triangle with sides 15, 36, 39, the Pythagorean theorem is
$39^2 = 15^2 + 36^2$

- In a right triangle, the line segment drawn from the right angle to the midpoint of the hypotenuse is equal to one-half of the hypothenuse:

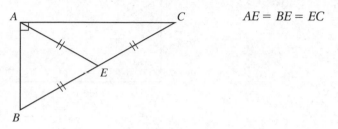

$AE = BE = EC$

SPECIAL RIGHT TRIANGLES

- In a **30-60-90 triangle**, the sides have the ratio of $x : x\sqrt{3} : 2x$.

Example: If $DF = 12$, DE must be half of that,
which is 6, and EF is $\sqrt{3}$ times DE,
which is $6\sqrt{3}$.

$$\frac{x}{ED} = \frac{x\sqrt{3}}{EF} = \frac{2x}{DF}$$

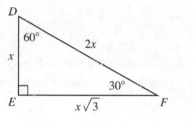

- In a **45-45-90 triangle**, the sides have the ratio of $x : x : x\sqrt{2}$.

Example: If $AC = 7$, BC must also be 7 and AB
is $\sqrt{2}$ times AC, which is $7\sqrt{2}$.

$$\frac{x}{AC} = \frac{x}{BC} = \frac{x\sqrt{2}}{AB}$$

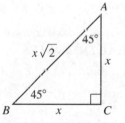

AREA OF A TRIANGLE

- The height (or altitude) of a triangle is a line segment from one of the vertices perpendicular to the side opposite that vertex. The side perpendicular to the height is called a base.

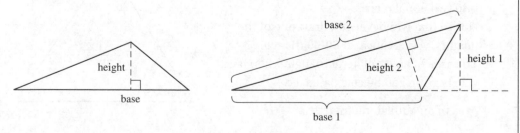

- The area and perimeter of a triangle can be calculated as follows:

Area $= \frac{1}{2} \times$ base $\times$ height $= \frac{1}{2}bh$

Perimeter $= a + b + c$

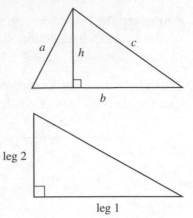

- The area of a right triangle becomes:

Area $= \frac{1}{2} \times \text{leg}_1 \times \text{leg}_2$

ISOSCELES TRIANGLES

- In an isosceles triangle, two sides and their opposite angles are equal.

- The altitude between the two equal sides also divides the angle of the vertex and the base into two equal pieces.

EQUILATERAL TRIANGLES

- In an equilateral triangle, all sides are equal and all angles are 60°.

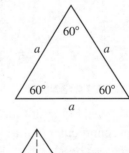

- Each height (altitude) of an equilateral triangle divides the vertex angle and the opposite side into two equal pieces.
- Notice that whenever you draw one of the heights, you create a 30-60-90 triangle.
- The area of an equilateral triangle when its side is given can be calculated as:

Area of an equilateral triangle $= \frac{\sqrt{3}}{4} a^2$

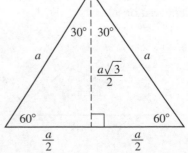

SIMILAR TRIANGLES

- **Similar triangles** have the same shape but not necessarily the same size. They have equal corresponding angles, and their corresponding sides are proportional.

Two triangles are similar if

$m\angle a = m\angle y$, $m\angle b = m\angle x$ and $m\angle c = m\angle z$

The corresponding sides of similar triangles are proportional:

$$\frac{AB}{DE} = \frac{AC}{DF} = \frac{BC}{EF}$$

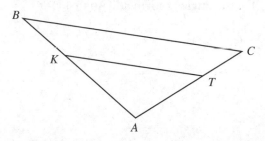

If KT is parallel to BC, then $\triangle AKT$ is similar to $\triangle ABC$.

If $\triangle AKT \sim \triangle ABC$,

$$\frac{AK}{AB} = \frac{AT}{AC} = \frac{KT}{BC}$$

> **NOTE**
>
> If a line segment parallel to one of the sides is drawn inside a triangle, the two triangles become similar.

SAMPLE PROBLEMS

EXAMPLE 1

What is the degree measure of one of the exterior angles of a regular 9-gon?

(A) 140°

(B) 120°

(C) 80°

(D) 60°

(E) 40°

The sum of the interior angles of a 9-gon can be found by the formula $(n - 2) \cdot 180° = (9 - 2) \cdot 180° = 7 \cdot 180°$

Since it is a regular 9-gon, find the degree measure of one of the interior angles by dividing the sum by 9: $7 \cdot \frac{180°}{9} = 140°$.

Finally, since an exterior angle and an interior angle at the same vertex always add up to 180°, the degree measure of an exterior angle is $180° - 140° = 40°$.

Alternatively, remember that the sum of all exterior angles of any polygon equals 360°. Therefore one exterior angle of a regular 9-gon is $\frac{360°}{9} = 40°$.

The answer is (E).

EXAMPLE 2

Based on the figure shown, 180° is how much greater than $7x$?

(A) 105°

(B) 75°

(C) 70°

(D) 15°

(E) 10°

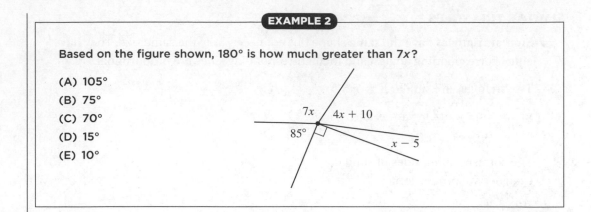

To find the value of x, set up an equation where all of the angles shown add up to 360°.

$$7x + 4x + 10 + x - 5 + 90 + 85 = 360$$
$$12x + 180 = 360$$
$$12x = 180$$
$$x = 15$$

This means $7x = 7 \cdot 15 = 105$.

180 is 75 more than 105 ($180 - 105 = 75$).

The answer is (B).

EXAMPLE 3

The height of a triangle equals $(t + 4)$, and its base equals $(t - 3)$. If the area of this triangle is 9 square inches, what is the value of t?

(A) 4

(B) 5

(C) 6

(D) 7

(E) 8

The area of a triangle is found by $\frac{1}{2} \cdot$ base $\cdot$ height, so $\frac{1}{2}(t + 4)(t - 3) = 9$.

$\frac{1}{2}(t + 4)(t - 3) = 9$	Multiply both sides by 2.
$(t + 4)(t - 3) = 18$	Distribute.
$t^2 + t - 12 = 18$	Subtract 18 from both sides.
$t^2 + t - 30 = 0$	Factor.
$(t + 6)(t - 5) = 0$	
$t + 6 = 0, \quad t - 5 = 0$	Since the sides cannot be negative, $t = 5$.
$t = -6 \qquad t = 5$	

The answer is (B).

1. What is the degree measure of angle C?

    (A) 40°
    (B) 50°
    (C) 60°
    (D) 70°
    (E) 80°

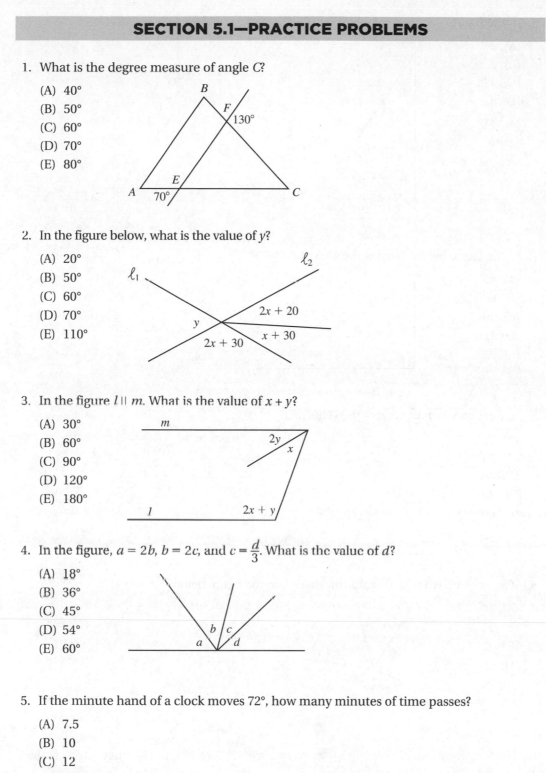

2. In the figure below, what is the value of y?

    (A) 20°
    (B) 50°
    (C) 60°
    (D) 70°
    (E) 110°

3. In the figure $l \parallel m$. What is the value of $x + y$?

    (A) 30°
    (B) 60°
    (C) 90°
    (D) 120°
    (E) 180°

4. In the figure, $a = 2b$, $b = 2c$, and $c = \dfrac{d}{3}$. What is the value of d?

    (A) 18°
    (B) 36°
    (C) 45°
    (D) 54°
    (E) 60°

5. If the minute hand of a clock moves 72°, how many minutes of time passes?

    (A) 7.5
    (B) 10
    (C) 12
    (D) 15
    (E) 45

6. In the figure, $AB = 4$. What is BC?

(A) $2\sqrt{2}$
(B) $2\sqrt{6}$
(C) $4\sqrt{2}$
(D) 8
(E) $4\sqrt{6}$

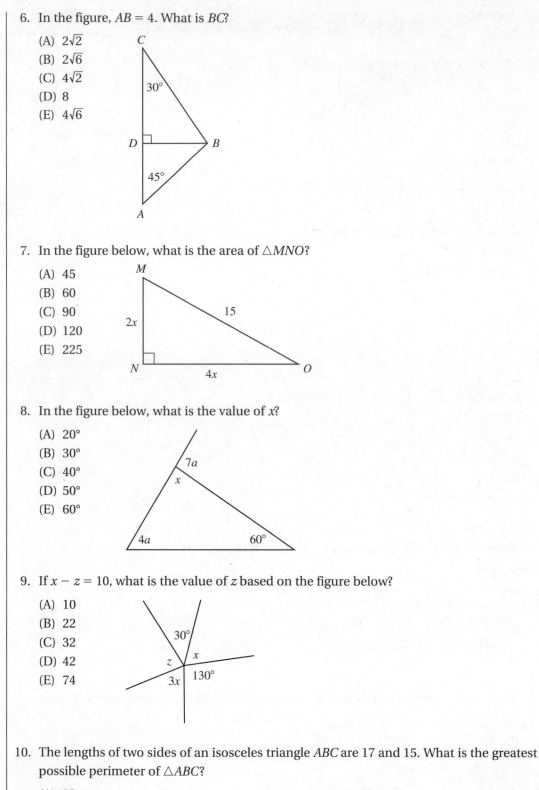

7. In the figure below, what is the area of $\triangle MNO$?

(A) 45
(B) 60
(C) 90
(D) 120
(E) 225

8. In the figure below, what is the value of x?

(A) 20°
(B) 30°
(C) 40°
(D) 50°
(E) 60°

9. If $x - z = 10$, what is the value of z based on the figure below?

(A) 10
(B) 22
(C) 32
(D) 42
(E) 74

10. The lengths of two sides of an isosceles triangle ABC are 17 and 15. What is the greatest possible perimeter of $\triangle ABC$?

(A) 32
(B) 35
(C) 37
(D) 39
(E) 49

11. In right triangle *DEF*, *E* is the right angle. If $DF = x + y$ and $EF = x - y$, what is the length of *DE* in terms of *x* and *y*?

(A) $\sqrt{2xy}$

(B) $2\sqrt{xy}$

(C) $2xy$

(D) xy

(E) $4xy$

12. In the figure below, which one of the sides is the longest?

(A) *AB*

(B) *AC*

(C) *BD*

(D) *BC*

(E) *CD*

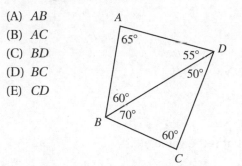

13. Points *P*, *Q*, and *R* are on the same line in that order. Point *S* is a point that is not on the same line and is equidistant from *Q* and *R*. If m∠*SQR* = 75° and m∠*PSQ* = 20°, what is the measure of ∠*SPQ*?

(A) 55°

(B) 50°

(C) 45°

(D) 40°

(E) 35°

14. Square *ABCD* is divided into 16 equal squares. If *AB* = 3, what is the area of the shaded triangle?

(A) 5

(B) $\frac{25}{16}$

(C) $\frac{45}{16}$

(D) $\frac{15}{4}$

(E) $\frac{45}{4}$

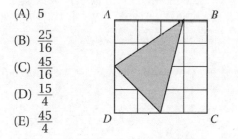

15. The figure shows a dial that is a regular hexagon. The arrow in the middle turns clockwise 150° every 20 seconds. Between which two numbers will it stop after 2 minutes and 10 seconds, if it starts by pointing to 1?

(A) 1 and 2

(B) 2 and 3

(C) 3 and 4

(D) 4 and 5

(E) 5 and 0

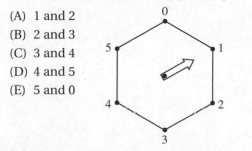

16. In the figure shown below, *a*, *b*, and *c* are external angles. If $235° < b + c < 275°$ and *a*, *b*, and *c* are integers, what is the greatest possible value of m∠*BAC*?

    (A) 86°
    (B) 88°
    (C) 90°
    (D) 92°
    (E) 94°

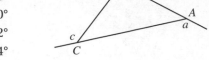

17. In triangle *PQR* m∠*P* is greater than 90°. If *PQ* = 3 and *PR* = 5, what is the sum of all possible integer values of *RQ*?

    (A) 12
    (B) 13
    (C) 14
    (D) 17
    (E) 19

18. In the figure below, m∠*ABC* = m∠*CBD*. What is the value of *x*?

    (A) 18°
    (B) 20°
    (C) 24°
    (D) 36°
    (E) 42°

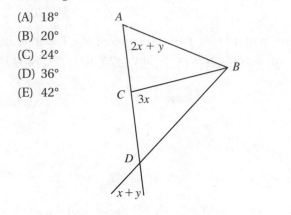

19. △*ABC* is a right triangle where m∠*C* = 90°. If the area of △*ABC* is 30 and $a - b = 7$, what is the length of the hypotenuse?

    (A) 11
    (B) 13
    (C) $10\sqrt{2}$
    (D) $10\sqrt{3}$
    (E) 19

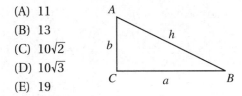

20. △*ABC* is an isosceles right triangle (*AB* = *AC*). If *AD* = 3 and *DC* = 8, how long is *BE*?

    (A) $5\sqrt{2}$
    (B) 7
    (C) $7\sqrt{2}$
    (D) 9
    (E) $8\sqrt{2}$

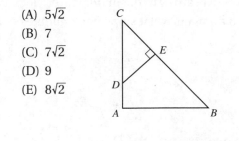

21. The sides of a triangle are $4k$, $5k$, and 6. Which of the following must be true for k?

    (A) $1 < k < 9$

    (B) $\frac{2}{3} < k < 9$

    (C) $\frac{2}{3} < k < 6$

    (D) $2 < k < 8$

    (E) $3 < k < 9$

22. Triangle MKT is an isosceles triangle. Based on the figure below, what is the ratio $\frac{y}{x} = ?$

    (A) $\sqrt{2}$

    (B) $\sqrt{3}$

    (C) 2

    (D) $2\sqrt{2}$

    (E) $3\sqrt{2}$

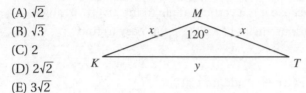

23. In the figure below, MNK and PQR are equilateral triangles. If the sum of their perimeters is 36 and the shaded area is $21\sqrt{3}$, how much longer is one side of MNK than one side of PQR?

    (A) 6

    (B) 7

    (C) 8

    (D) $6\sqrt{3}$

    (E) $7\sqrt{3}$

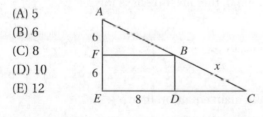

24. AEC is a right triangle, and $FBDE$ is a rectangle. Based on the lengths of the sides shown in the figure, what is the value of x?

    (A) 5

    (B) 6

    (C) 8

    (D) 10

    (E) 12

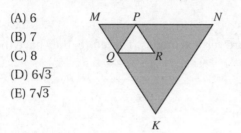

25. Based on the figure, what is the value of a in terms of b?

    (A) $\frac{b}{3}$

    (B) $30 - \frac{b}{3}$

    (C) $30 - b$

    (D) $90 - \frac{b}{3}$

    (E) $30 + \frac{b}{3}$

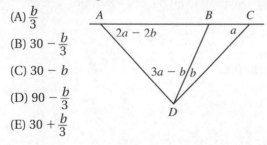

26. In the figure below, with $\ell_1 \parallel \ell_2$, what is the value of x?

    (A) 30°

    (B) 35°

    (C) 40°

    (D) 45°

    (E) 50°

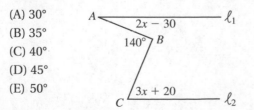

1	2	3	4	5	6	7	8	9	10	11	12	13	14	15	16	17	18	19	20
C	E	B	D	C	C	A	C	C	E	B	E	A	C	E	E	B	D	B	C

21	22	23	24	25	26
C	B	B	D	E	A

1. **C** $\angle EFC$ is equal to 50° because $\angle EFC$ and the 130° angle are on a straight line
   $(180 - 130 = 50)$. $\angle FEC$ is 70° because it is a vertical angle to the given 70° angle.
   Now that you have two of the interior angles of $\triangle FEC$, $\angle C$ is easy to find.
   $m\angle C = 180 - 50 - 70 = 60$.

2. **E** Notice that $2x + 20$, $x + 30$, and $2x + 30$ add up to 180°.

   $$2x + 20 + x + 30 + 2x + 30 = 180$$
   $$5x + 80 = 180$$
   $$5x = 100$$
   $$x = 20$$

   Angle y and the sum of $2x + 20$ and $x + 30$ are vertical angles. Therefore, they have the
   same measurement.

   $2x + 20 + x + 30 = 40 + 20 + 20 + 30 = 110$.
   So y also equals 110.

3. **B** Since ℓ is parallel to m, $x + 2y$ on top and $2x + y$ on the bottom add up to 180°.
   Notice that this is the only equation that can be written. Therefore it is not possible to
   solve for x and y separately. On the other hand, the question asks for $x + y$.

   $$x + 2y + 2x + y = 180°$$
   $$3x + 3y = 180° \qquad \text{Divide both sides by 3.}$$
   $$x + y = 60°$$

4. **D** Since all angles add up to 180°, write the main equation first:
   $a + b + c + d = 180°$.
   Each angle needs to be written in terms of one single angle. For example, pick c:

$b = 2c$	Given.
$\dfrac{d}{3} = c$	This means $d = 3c$.
$a = 2b$	Since $b = 2c$, replace $2c$ for b.
$a = 2 \cdot 2c = 4c$	

   Replace all into the equation above:

   $$4c + 2c + c + 3c = 180°$$
   $$10c = 180$$
   $$c = 18$$
   $$d = 3c = 3 \cdot 18 = 54$$

5. **C** The minute hand sweeps 360° in 60 minutes. Set up a proportion:

$$\frac{72°}{360°} = \frac{x}{60 \text{ minutes}}$$

$$x = \frac{60 \text{ minutes} \cdot 72°}{360°} = \frac{72}{6} = 12 \text{ minutes}$$

6. **C** In order to find BC, you need to find one of the sides of $\triangle BDC$ needed. Since $\triangle ABD$ and $\triangle BDC$ share DB, find DB using the given length AB.

$\triangle ABD$ is a 45-45-90 triangle. Therefore, the ratios of the sides are $AD : DB : AB = x : x : x\sqrt{2}$. This means:

$$\frac{DB}{1} = \frac{AB}{\sqrt{2}}$$

$$\frac{DB}{1} = \frac{4}{\sqrt{2}}$$

$$DB = \frac{4 \cdot \sqrt{2}}{\sqrt{2} \cdot \sqrt{2}} = \frac{4\sqrt{2}}{2} = 2\sqrt{2}$$

$\triangle CDB$ is a 30-60-90 triangle. The ratio of its sides are $DB : CD : BC = x : x\sqrt{3} : 2x$, This means:

$$\frac{DB}{1} = \frac{BC}{2}$$

$$BC = 2 \cdot DB = 2 \cdot \sqrt{2} = 4\sqrt{2}$$

7. **A** The area of a triangle is $\frac{1}{2} \cdot$ base $\cdot$ height. We are looking for $\frac{1}{2} \cdot (4x) \cdot (2x) = 4x^2$. Notice that we do not need to find x but only $4x^2$.

Use the Pythagorean theorem:

$$15^2 = (4x)^2 + (2x)^2$$

$$225 = 16x^2 + 4x^2 = 20x^2$$

Since we need $4x^2$, divide both sides of the equation by 5:

$$\frac{225}{5} = \frac{20x^2}{5}$$

$$45 = 4x^2 = \text{Area}$$

8. **C** In any triangle, an exterior angle is the sum of the two remote interior angles.

$$7a = 4a + 60$$

$$3a = 60$$

$$a = 20$$

One of the interior angles of the triangle is

$$4a = 4 \cdot 20 = 80$$

The interior angles add up to 180.

$$x + 80 + 60 = 180$$

$$x = 40$$

9. **C** From the figure, $z + 30 + x + 130 + 3x = 360$. Add the like terms and simplify to get $z + 4x + 160 = 360$.

Subtract 160 from both sides to get, $z + 4x = 200$. Since the question provides $x - z = 10$, solve the two equations simultaneously using elimination:

$$
\begin{array}{r}
z + 4x = 200 \\
+ \quad x - z = 10 \\
\hline
5x = 210 \\
x = 42
\end{array}
$$

If $x = 42$, $42 - z = 10$. So $z = 32$.

10. **E** There are two possible isosceles triangles: 17-17-15 and 15-15-17. The one with the longer perimeter is the first one: $17 + 17 + 15 = 49$.

11. **B** Use the Pythagorean theorem: $DF^2 = EF^2 + DE^2$.

$$(x + y)^2 = (x + y)^2 + DE^2$$
$$x^2 + 2xy + y^2 = x^2 - 2xy + y^2 + DE^2$$

Subtract x^2 and y^2 from each side.

$2xy = -2xy + DE^2$ Add $2xy$ to each side.
$4xy = DE^2$ Take the square root of both sides.
$DE = 2\sqrt{xy}$

12. **E** In triangle *ABD*, the longest side is *BD* because it is across from the largest angle.

$BD > AD > AB$ since $65 > 60 > 55$.

In triangle *BCD*, $CD > BD > BC$ since $70 > 60 > 50$. Therefore *CD* must be the longest side.

13. **A** First, sketch the points as described:

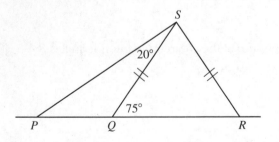

Notice that 75° is an external angle of triangle *PSQ*. Since an external angle is the sum of the two opposite internal angles,

$75° = 20° + \mathrm{m}\angle SPQ$. Therefore $\mathrm{m}\angle SPQ = 55°$.

14. **C**

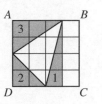

In this case, it is easier to count the number of squares that are not shaded first.

There are 4 squares to the right side. Area 1 is half of 4 squares, so it equals 2 squares. Area 2 is also half of 4 squares, so it equals 2 squares. Area 3 is half of 6 squares, so it equals 3 squares. In total, $4 + 2 + 2 + 3 = 11$ squares are not shaded in the original problem. This means $16 - 11 = 5$ squares are shaded.

Each small square has a side of $\frac{3}{4}$ and therefore an area of $\left(\frac{3}{4}\right)^2 = \frac{9}{16}$. Since 5 squares were shaded, the total area equals $5 \cdot \frac{9}{16} = \frac{45}{16}$.

15. **E** First convert minutes into seconds since the turning speed is given in degrees per second: 2 minutes and 10 seconds = $2 \cdot 60 + 10 = 130$ seconds. Then set up a proportion to find how many degrees the arrow will turn:

$$\frac{\text{Degrees}}{\text{Seconds}} = \frac{150°}{20 \text{ seconds}} = \frac{x°}{130 \text{ seconds}}$$

Cross multiply $20x = 130 \cdot 150$. So $x = 975°$.

Every 360° is a complete turn, and 975° implies that there were two complete turns ($360 \cdot 2 = 720°$) plus some more. The additional turn after the arrow comes back to number 1 is $975° - 720° = 255°$

The angle between each number is 60° because this is a regular hexagon and $\frac{360°}{6} = 60°$.

Finally, you can find how many numbers the arrows passes by dividing 255 by 60. 255 is more than $60 \cdot 4 = 240$. So the arrow will stop after 5 but before 0.

16. **E** The external angles of a triangle add up to 360°. Therefore $a + b + c = 360°$.

If we are trying to maximize the degree measure of $\angle BAC$, we will need to minimize the value of a since they are on a line and add up to 180°. To minimize a, pick the largest integer value for $b + c$, which is 274°. Substitute $b + c = 274°$ in $a + b + c = 360°$ and find $a = 360° - 274° = 86°$. Since $a + m\angle BAC = 180°$, $m\angle BAC = 180° - 86° = 94°$.

17. **B** Compare this triangle to a right triangle with legs equal to 3 and 5. The hypotenuse in that case would be $n = \sqrt{3^2 + 5^2} = \sqrt{34}$. $\sqrt{34}$ is slightly less than 6 since $6^2 = 36$.

In PQR, $m\angle P$ is greater than 90°, which means that side RQ must be longer than $\sqrt{34}$. The first integer value greater than $\sqrt{34}$ is 6. There's also an upper limit to the values that RQ can take. Since the other two sides are 3 and 5, RQ must be less than 8. Therefore RQ can only be 6 or 7 since it has to be an integer, $6 + 7 = 13$.

18. **D** Let $m\angle ABC = k$, which also means that $m\angle CBD = k$. Write the sum of the interior angles of triangles ABC and CBD. $m\angle ACB = 180 - 3x$ and $m\angle CBD = x + y$.

For $\triangle ABC$: $2x + y + 180 - 3x + k = 180$
For $\triangle CBD$: $3x + x + y + k = 180°$
Set these equations equal to each other.

$$2x + y + 180 - 3x + k = 3x + x + y + k$$

Subtract k and y from both sides, and simplify.

$$2x - 3x + 180 = 3x + x$$
$$-x + 180 = 4x$$
$$180° = 5x$$
$$36° = x$$

19. **B** The area of $\triangle ABC$ can be written as $\frac{a \cdot b}{2} = 30$, so $a \cdot b = 60$. If $a - b = 7$, you can use substitution to solve for a and b.

$a - b = 7$ means $a = b + 7$, which can be substituted into $a \cdot b = 60$.

$$(b + 7)b = 60$$
$$b^2 + 7b = 60$$
$$b^2 + 7b - 60 = 0.$$

Factor this equation as $(b + 2)(b - 5) = 0$. $b = -12$ or $b = 5$.
Since b is a length, $b = 5$. Use $a - b = 7$, to find that a must be 12. Use the Pythagorean theorem to find $h = \sqrt{12^2 + 5^2} = 13$.

Alternatively, start writing down $h^2 + a^2 + b^2$, which means we are trying to get to $a^2 + b^2$. Square both sides in $(a - b) = 7$ to yield $a^2 - 2ab + b^2 = 49$. Since we know that $a \cdot b = 60$ from the area equation, substitute it into our equation.

$$a^2 - 2ab + b^2 = 49$$
$$a^2 - 2 \cdot 60 + b^2 = 49$$
$$a^2 + b^2 = 49 + 120$$
$$a^2 + b^2 = 169$$
$$h = \sqrt{a^2 + b^2} = \sqrt{169} = 13$$

20. **C** If $AD = 3$ and $DC = 8$, one leg of the isosceles right triangle is 11. Also we know that $m\angle C = m\angle B = 45°$ in an isosceles right triangle (45-45-90 triangle). Using the Pythagorean theorem or the ratios of sides of a 45-45-90 triangle, we can calculate BC as $11\sqrt{2}$.

Triangle CDE is also a 45-45-90 triangle. From the ratios of the sides of this special triangle we can write:

$$\frac{CD}{\sqrt{2}} = \frac{CE}{1}$$

$$\frac{8}{\sqrt{12}} = \frac{CE}{1}$$

$$CE = \frac{8}{\sqrt{12}} \cdot \frac{\sqrt{2}}{\sqrt{2}} = \frac{8\sqrt{2}}{2} = 4\sqrt{2}$$

Since $CE = 4\sqrt{2}$, BE becomes $11\sqrt{2} - 4\sqrt{2} = 7\sqrt{2}$.

21. **C** In every triangle, one side is less than the sum and greater than the difference of the other sides. Therefore $6 < 4k + 5k$ and $6 > 5k - 4k$. $6 < 9k$ and $\frac{6}{9} < k$, which means $\frac{2}{3} < k$ and $k < 6$. The two inequalities can be combined as follows:

$$\frac{2}{3} < k < 6$$

22. **B** Draw the attitude from vertex M to point A on side KT. This divides angle M and side KT into two equal pieces. From the 30-60-90 triangle ratios:

$$\frac{MT}{2} = \frac{AT}{\sqrt{3}}$$

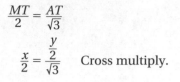

$$\frac{x}{2} = \frac{\frac{y}{2}}{\sqrt{3}} \quad \text{Cross multiply.}$$

Since $x\sqrt{3} = y$, then $\frac{y}{x}$ must be $\sqrt{3}$.

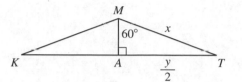

23. **B** Let the length of the sides of MNK be x, and let the length of the sides of PQR be y.

If their perimeters add up to 36, $3x + 3y = 36$ and $x + y = 12$. The shaded area represents the difference of their areas. Since the area of an equilateral triangle is $\frac{a^2\sqrt{3}}{4}$,

$$\frac{\sqrt{3}}{4}x^2 - \frac{\sqrt{3}}{4}y^2 = 21\sqrt{3}$$

Factor out $\frac{\sqrt{3}}{4}$.

$$\frac{\sqrt{3}}{4}(x^2 - y^2) = 21\sqrt{3}$$

Multiply each side by $\frac{4}{\sqrt{3}}$.

$$x^2 - y^2 = 84$$

Since $x^2 - y^2 = (x - y)(x + y) = 84$

and $x + y = 12$, substituting gives us $(x - y) \cdot 12 = 84$ and $x - y = 7$.

24. **D** Since $FBDE$ is a rectangle, all of its angles are 90°. That makes $\triangle AFB$ and $\triangle BDC$ right triangles. Additionally, $FB = 8$ and $BD = 6$.

Using the Pythagorean theorem for $\triangle AFB$ gives us $AB^2 = 6^2 + 8^2 = 100$. So $AB = 10$.

$\triangle AFB$ is similar to $\triangle BDC$ since $m\angle A = m\angle CBD$ and $m\angle ABF = m\angle C$

$$\frac{AF}{BD} = \frac{AB}{BC}$$

$$\frac{6}{6} = \frac{10}{x}$$

$$x = 10$$

25. **E** The interior angles of triangles add up to 180°.

In $\triangle ADC$,
$$2a - 2b + 3a - b + a + b = 180°$$
$$6a - 2b = 180°$$
$$3a - b = 90°$$
$$3a = 90° + b$$
$$a = 30° + \frac{b}{3}$$

26. **A** Draw a line through B that is parallel to ℓ_1 and ℓ_2.

$\angle ABX = 2x - 30$

$\angle CBX = 3x + 10$

Since $\ell_1 \parallel \ell_2 \parallel \ell_3$,

$$2x - 30 + 3x + 20 = 140$$
$$5x - 10 = 140$$
$$5x = 150$$
$$x = 30°$$

SECTION 5.2: QUADRILATERALS, CIRCLES, AND SOLIDS
Quadrilaterals

- A quadrilateral is a polygon with four sides.

PARALLELOGRAMS

- A parallelogram is a quadrilateral with two sets of parallel sides.
- The opposite sides of a parallelogram are parallel and have equal lengths.

$AB = DC$ and $AD = BC$

$AB \parallel DC$ and $AD \parallel BC$

- Diagonals of a parallelogram bisect each other.

$DM = MB$ and $AM = MC$

- The area of a parallelogram is base × height = *bh*.

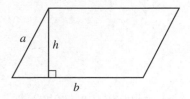

- The perimeter of a parallelogram = $a + a + b + b = 2(a + b)$

RECTANGLES

- A parallelogram with right angles is called a rectangle. All properties of parallelograms apply to rectangles as well.
- *DB* and *AC* are diagonals, and they are equal in length. By using the Pythagorean theorem, we can find the length as $DB^2 = a^2 + b^2$ since *ABD* is a right triangle.

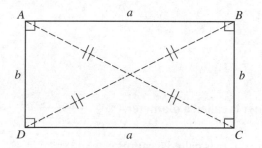

- The area of a rectangle is length × width = $a \cdot b$.
- The perimeter of a rectangle is $a + a + b + b = 2(a + b)$.

SQUARE

- A square is a rectangle with all sides equal to each other.
- The area of a square is length × width = $a \cdot a = a^2$.

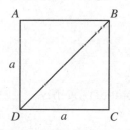

- The perimeter of a square is $a + a + a + a = 4a$.
- The diagonal of a square, such as *DB*, equals $a\sqrt{2}$.

TRAPEZOIDS

- A trapezoid is a quadrilateral with only two parallel sides.
  AB ∥ *DC*

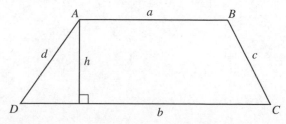

- The area of a trapezoid is the product of its height and the average of the two parallel sides (bases): $\frac{a+b}{2} \cdot h = \frac{1}{2}(a+b)h$.
- The perimeter of a trapezoid is $a + b + c + d$.

CIRCLES

- A circle is a set of points in a plane equidistant from a fixed point (the center).
- Any line segment that has its endpoints on the circle is called a **chord**.
  DE and AB are chords.

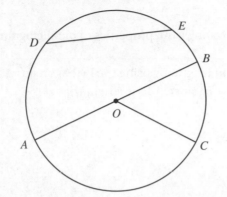

- Any chord that passes through the center is called a **diameter**.
  AB is a diameter.
- Any line segment from the center to the circle is called a **radius**.
  OA, OB, and OC are radii (plural for radius.)
- The length of a diameter is twice the length of a radius.
- Any line that has only one point common with a circle is called a **tangent**.
  ℓ_1 is tangent to the circle at K. K is called the point of tangency.

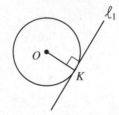

- A tangent line is perpendicular to the radius drawn to the point of tangency.
- The area of a circle is πr^2, where r is the radius.
- The circumference of a circle is $2\pi r = \pi d$, where r is the radius and d is the diameter.

Examples: The area of the circle below is $\pi \cdot r^2 = \pi \cdot 12^2 = 144\pi$.

The circumference of the circle is $2 \cdot \pi \cdot r = 24\pi$.

PI

π is approximately 3.14 or $\frac{22}{7}$.

THE RATIO OF THE AREAS

The ratio of the areas of two circles is the square of the ratio of their radii.

$$\frac{A_1}{A_2} = \frac{r_1^2}{r_2^2}$$

Example: The ratio of the diameters of two circular rugs is 2 : 3. What is the ratio of their areas?

If the ratio of their diameters is 2 : 3, the ratio of their radii is also 2 : 3.

Based on the previous information, the ratio of their areas is $\left(\frac{2}{3}\right)^2 = \frac{4}{9}$.

To check, compare the actual areas, $\frac{A_1}{A_2} = \frac{\pi r_1^2}{\pi r_2^2} = \frac{r_1^2}{r_2^2}$.

- An angle with a vertex at the center of a circle is called a **central** angle. α in the figure is a central angle.

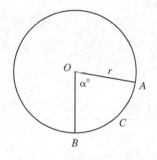

- An **arc** is a piece of a circle.
- The degree measure of arc ACB is equal to the central angle AOB (or α).
- The degree measure of the entire circle is 360°.
- The ratio of the length of arc ACB to the circumference of the circle is $\frac{\alpha}{360°}$. Therefore, the length of arc ACB can be calculated as:

Length of arc $ACB = 2\pi r \cdot \frac{\alpha}{360}$

Example: What is the length of an arc with a central angle of 120° in a circle with a radius of 8 inches?

Arc length $= 2\pi \cdot 8 \cdot \frac{120}{360} = \frac{16\pi}{3}$

- The ratio of the area of the sector (slice) AOB to the area of the circle is $\frac{\alpha}{360°}$. Therefore, the area of sector AOB is calculated as:

Area of sector $AOB = \pi \cdot r^2 \cdot \frac{\alpha}{360}$

Example: What is the area of a sector with a central angle of 120° in a circle with a radius of 8 inches?

Sector area $= \pi \cdot 8^2 \cdot \frac{120}{360} = \pi \cdot 64 \cdot \frac{1}{3} = \frac{64\pi}{3}$

- An angle whose vertex is on the circle and sides are chords of the circle is called an **inscribed angle**.
- If a triangle has three vertices on the circle (inscribed in a circle) and one of its sides is a diameter, it is a right triangle. In other words, if an inscribed angle's sides intersect the circle at the endpoints of a diameter, it is a right angle.

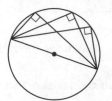

- If an inscribed angle and a central angle intersect the circle at the same points, the degree measure of the inscribed angle is half of the central angle (or half of the arc that the inscribed angle cuts from the circle).

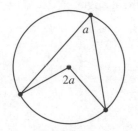

- If each vertex of a polygon lies on a circle, the **polygon** is **inscribed** in the circle or the **circle is circumscribed** about the poygon.

- If each side of a polygon is tangent to a circle, the **polygon is circumscribed** about the circle or the **circle is inscribed** in the polygon.

Solids

RECTANGULAR PRISMS

- A **rectangular prism** is a three-dimensional box with 6 rectangular faces. **Edges** are the lines where the faces meet, such as lines *FB* or *GH*. Each point where the edges intersect is called a **vertex**, such as points *C* or *D*.

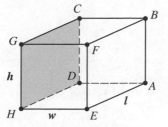

- A rectangular solid has 6 faces and 8 vertices.
- To find the volume of a rectangular solid, simply multiply the lengths of the three dimensions.
- Volume of a rectangular solid = Length × Width × Height = $l \times w \times h$
- To find the surface area of a rectangular solid, find the sum of all surface areas individually and add them together. Notice that there are three pairs of equal rectangular faces. For example, the area of the gray face above (rectangle *CDHG*) is $h \times l$. Rectangle *AEFB* has the same area.
- Total Surface Area of a Rectangular Solid = $2hw + 2wl + 2hl = 2(hw + wl + hl)$

CUBES

- A rectangular solid with all edges equal is called a **cube**.
- Volume of a Cube = Length × Width × Height = $a \times a \times a = a^3$

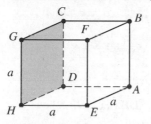

- Total Surface Area of a Cube = $6(a \times a) = 6a^2$

CYLINDERS

- A right circular **cylinder** has identical circular bases. Its height is perpendicular to its faces.
- Volume of a Cylinder = Base Area × Height = $\pi \times r^2 \times h = \pi r^2 h$

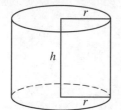

- To find the surface area of a cylinder, find the areas of the bases and the surface. Then add them together. Notice that the curved side surface is actually a rectangle with height equal to the height of the cylinder and width equal to the circumference of the base.
- Total Surface Area of a Cylinder = $2(\pi r^2) + 2\pi rh$

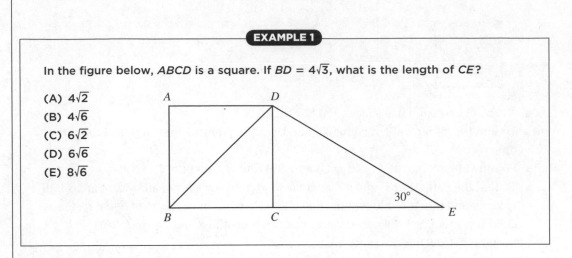

EXAMPLE 1

In the figure below, *ABCD* is a square. If $BD = 4\sqrt{3}$, what is the length of *CE*?

(A) $4\sqrt{2}$

(B) $4\sqrt{6}$

(C) $6\sqrt{2}$

(D) $6\sqrt{6}$

(E) $8\sqrt{6}$

In order to find *CE*, one of the sides of $\triangle CDE$ is needed. Since square *ABCD* and $\triangle CDE$ share *CD*, start with finding *CD* using the given length *BD*.

When a diagonal of a square is drawn, it creates two special triangles with 45-45-90 angles. Therefore, the ratios of the sides of $\triangle BCD$ are $BC : CD : BD = x : x : x\sqrt{2}$.

$$\frac{CD}{x} = \frac{BD}{x\sqrt{2}}$$

$$\frac{CD}{x} = \frac{4\sqrt{3}}{x\sqrt{2}}$$

$$CD = \frac{4\sqrt{3}}{\sqrt{2}}$$

$\triangle CDE$ is a 30-60-90 triangle. The ratios of its sides are $CD : CE : DE = x : x\sqrt{3} : 2x$.

$$\frac{CD}{x} = \frac{CE}{x\sqrt{3}}$$

$$CE = \sqrt{3} \cdot CD$$

$$CE = \sqrt{3} \cdot \frac{4\sqrt{3}}{\sqrt{2}} = \frac{12}{\sqrt{2}} = \frac{12 \cdot \sqrt{2}}{\sqrt{2} \cdot \sqrt{2}} = \frac{12\sqrt{2}}{2} = 6\sqrt{2}$$

The answer is (C).

In the figure below, *EAC* is an equilateral triangle and *KLWM* is a square. What is the ratio of the area of △*LWC* to the area of △*EKL*?

(A) $\frac{\sqrt{3}}{3}$

(B) $\frac{\sqrt{6}}{3}$

(C) $\frac{1}{3}$

(D) $\frac{2}{3}$

(E) $\sqrt{6}$

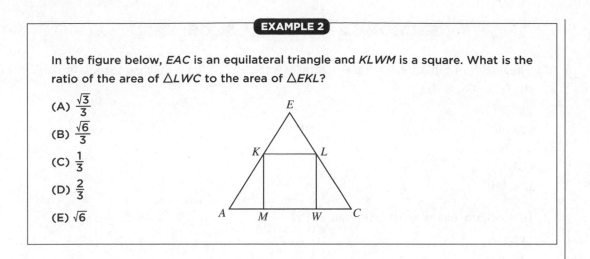

Let one side of the square be x. So $KL = LW = WM = MK = KE = LE = x$.

$m\angle C = 60°$ since △*EAC* is equilateral, and $m\angle LWC = 90°$ since *KLWM* is a square.

△*LWC* is a 30-60-90 triangle.

$$\frac{LW}{\sqrt{3}} = \frac{WC}{1}$$

$$\frac{x}{\sqrt{3}} = \frac{WC}{1}$$

$$WC = \frac{x}{\sqrt{3}} = \frac{x \cdot \sqrt{3}}{\sqrt{3} \cdot \sqrt{3}} = \frac{x\sqrt{3}}{3}$$

$$\frac{\text{Area of } \triangle LWC}{\text{Area of } \triangle EKL} = \frac{\frac{1}{2} \cdot x \cdot \frac{x\sqrt{3}}{3}}{x^2 \cdot \frac{\sqrt{3}}{4}} = \frac{\frac{x^2\sqrt{3}}{6}}{\frac{x^2\sqrt{3}}{4}} = \frac{x^2\sqrt{3}}{6} \cdot \frac{4}{x^2\sqrt{3}} = \frac{2}{3}$$

The answer is (D).

A cylindrical container has a diameter of 14 m and is half filled with water. A cement block in the shape of a rectangular prism is dropped into the water tank and is completely submerged. If the sides of the cement block are 11 m × 7 m × 9 m, how much does the water level in the tank rise? Use $\pi \approx \frac{22}{7}$.

(A) 2　　　(B) $\frac{9}{4}$　　　(C) 3　　　(D) $\frac{9}{2}$　　　(E) 4

If the cement block is completely submerged, the volume of water that would rise is $11 \cdot 7 \cdot 9 \text{ m}^3$.

To find the height increase, think of this increase as the volume of a cylinder with a base diameter of 14 m ($r = 7$).

$V = \pi r^2 h = \frac{22}{7} \cdot 7^2 \cdot h$　　　must equal　$11 \cdot 7 \cdot 9$.

$22 \cdot 7 \cdot h = 11 \cdot 7 \cdot 9$　　　Simplify.

$2h = 9$　　　Divide by 2.

$h = \frac{9}{2}$

The answer is (D).

1. The angles of a quadrilateral have a ratio of $2:3:6:7$. What is the degree measure of the largest angle?

   (A) 20°
   (B) 40°
   (C) 60°
   (D) 120°
   (E) 140°

2. In the figure below, what is the value of x?

   (A) 15°
   (B) 45°
   (C) 75°
   (D) 105°
   (E) 195°

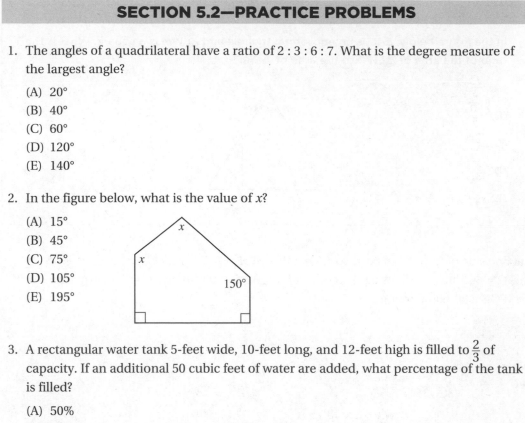

3. A rectangular water tank 5-feet wide, 10-feet long, and 12-feet high is filled to $\frac{2}{3}$ of capacity. If an additional 50 cubic feet of water are added, what percentage of the tank is filled?

   (A) 50%
   (B) 66.7%
   (C) 72%
   (D) 75%
   (E) 96%

4. If an arc with length 12π is $\frac{3}{4}$ of the circumference of a circle, what is the shortest distance between the endpoints of the arc?

   (A) 4
   (B) $4\sqrt{2}$
   (C) 8
   (D) $8\sqrt{2}$
   (E) $8\sqrt{3}$

5. The length of each side of square A is increased by 100% to make square B. If the length of each side of B is doubled to make square C, the area of A is what fraction of the sum of the areas of B and C?

   (A) $\frac{1}{4}$

   (B) $\frac{1}{8}$

   (C) $\frac{1}{16}$

   (D) $\frac{1}{20}$

   (E) $\frac{1}{64}$

6. In the figure below, all sides meet at right angles. Based on the sides given, what is the perimeter of the figure?

(A) 140
(B) 120
(C) 80
(D) 70
(E) 60

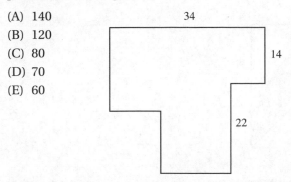

7. In the figure below, line *KT* is tangent to the circle at *K* and *M* is the center of the circle. What is the radius of the circle if *KT* = 15 and *MT* = 17?

(A) 2
(B) 5
(C) 6
(D) 8
(E) 9

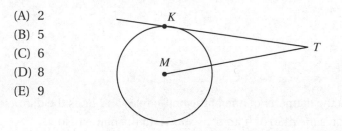

8. In the figure, *FD* equals *FB* and *FE* equals *FC*. If the degree measure of arc *CAB* is 100°, what is the measure of angle *x*?

(A) 140°
(B) 120°
(C) 80°
(D) 70°
(E) 60°

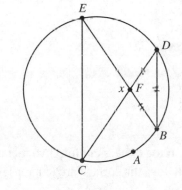

9. What is the area of the trapezoid shown below?

(A) $9\sqrt{3}$
(B) 27
(C) $27\sqrt{3}$
(D) 36
(E) $36\sqrt{3}$

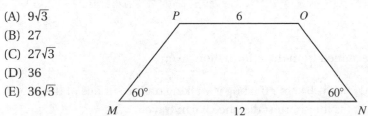

10. In the figure, $ABCD$ is a square. Arc BD is $\frac{1}{4}$ of a circle with its center at A. If $BC = 6$, what is the closest approximation of the area of the shaded region?

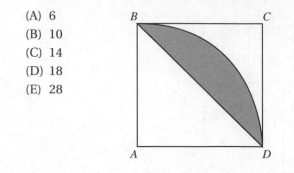

(A) 6
(B) 10
(C) 14
(D) 18
(E) 28

11. The dimensions of a rectangular prism are 12 cm, 8 cm, and 4 cm. How many cubes with edges 2 cm long can be fit into this prism?

(A) 8
(B) 10
(C) 12
(D) 24
(E) 48

12. ABC is a right triangle. AC is the diameter of one of the half-circles and BC is the diameter of the other half-circle. What is the ratio of A_1 to A_2 as shown in the figure below?

(A) $\frac{3}{2}$

(B) $\sqrt{2}$

(C) 2

(D) $\frac{5}{2}$

(E) 3

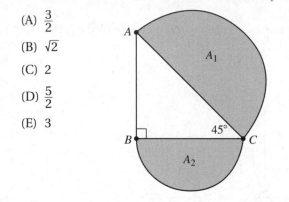

13. An edge of a solid cube is 8 inches long. If the cube is cut into two rectangular prisms parallel to one of its faces, by how much does the surface area increase?

(A) It stays the same
(B) 32
(C) 64
(D) 128
(E) It cannot be determined from the information given

14. The volume of a solid cube is 64 cm³. If a bug is walking on the surface of the cube from point A to point Z, what is the shortest distance it can travel?

(A) 8
(B) $4\sqrt{3}$
(C) $4 + 4\sqrt{2}$
(D) $4\sqrt{5}$
(E) $16\sqrt{2}$

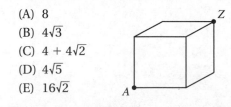

15. In the regular pentagon below, what is the m∠*BEC*?

(A) 15°
(B) 18°
(C) 24°
(D) 36°
(E) 48°

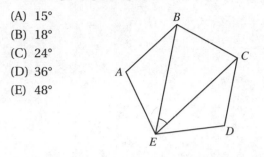

16. In the figure below, *ABC* is an equilateral triangle and *EFDC* is a parallelogram. If the perimeter of △*ABC* is 48, what is the perimeter of *EFDC*?

(A) 12
(B) 16
(C) 24
(D) 32
(E) 36

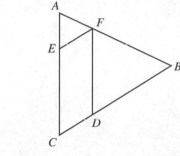

17. In the figure below, *ACEF* is a parallelogram and *BC* = *CD*. If m∠*F* = 40° and m∠*FGB* = 165°, what is the m∠*GBD*?

(A) 75°
(B) 85°
(C) 95°
(D) 105°
(E) 115°

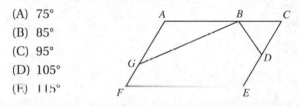

18. In the circle below, *O* is the center. Three times the length of arc *AB* equals twice the length of arc *CB*. Arc *CA* is twice as long as arc *AB*. What is the measure of angle *x*?

(A) 75°
(B) 80°
(C) 85°
(D) 90°
(E) 95°

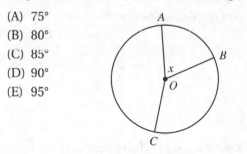

19. If *AD* = 12 and *AB* = 8, what is the area of the trapezoid *ABCD*?

(A) $22 + 6\sqrt{3}$
(B) $22\sqrt{3} + 6$
(C) $66 + 18\sqrt{3}$
(D) $66\sqrt{3} + 18$
(E) $66\sqrt{3} + 54$

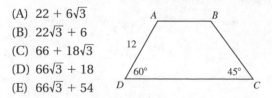

20. *ACEF* is a parallelogram where $BC = AB$ and $2 \cdot DE = CD$. If the area of $\triangle BCD = 12$, what is the area of *ABDEF*?

    (A) 96
    (B) 84
    (C) 72
    (D) 60
    (E) 48

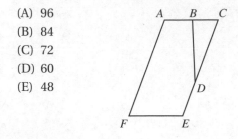

21. Five identical circles are tangent to each other and to the outer circle as shown. What is the ratio of the total area of the small circles to the area of the large circle?

    (A) $\dfrac{1}{9}$

    (B) $\dfrac{2}{9}$

    (C) $\dfrac{1}{3}$

    (D) $\dfrac{4}{9}$

    (E) $\dfrac{5}{9}$

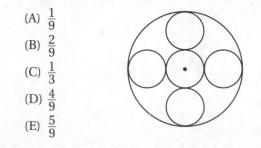

22. Two identical quarter-circles are drawn inside square *ABCD* as shown. If $AD = 2$, what is the shaded area?

    (A) $2\pi - 2$
    (B) $2\sqrt{2} - \pi$
    (C) $4\sqrt{2} - \pi$
    (D) $4 - \pi$
    (E) $4\sqrt{2} - \pi$

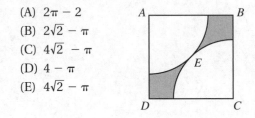

23. A rectangular sheet of aluminum has a length of 16 and a width of 3π. A manufacturer cuts the maximum possible number of identical circular disks with radii 2 from this sheet and discards the rest. What is the area of the discarded sheet?

    (A) 0
    (B) 2π
    (C) 3π
    (D) 8π
    (E) 16π

24. A cylindrical piece of wood will be cut into identical cubes. The diameter of the cylinder is $6\sqrt{2}$ ft and the height is 16 ft. If the edge of the cubes must be at least 4 ft, what is the greatest total volume of the resulting cubes?

    (A) 256 ft³
    (B) 375 ft³
    (C) 432 ft³
    (D) 459 ft³
    (E) 496 ft³

1	2	3	4	5	6	7	8	9	10	11	12	13	14	15	16	17	18	19	20
E	D	D	D	D	A	D	C	C	B	E	C	D	D	D	D	B	B	E	D

21	22	23	24
E	D	E	C

1. **E** Let the angles of the quadrilateral be A, B, C, and D. Then $\frac{A}{2} = \frac{B}{3} = \frac{C}{6} = \frac{D}{7} = k$. Express each angle in terms of k.

   $A = 2k$, $B = 3k$, $C = 6k$, and $D = 7k$. Since the sum of the interior angles of a quadrilateral is 360, $2k + 3k + 6k + 7k = 360$ and $18k = 360$.

   Solve for k to get $k = 20$. The largest angle is D, which is $7k = 7 \cdot 20 = 140°$.

2. **D** The sum of all interior angles of a pentagon can be found by using the formula

   $(n - 2) \cdot 180 = (5 - 2) \cdot 180 = 540$. Add all the angles of the pentagon and set them equal to 540.

   $$90 + 90 + 150 + x + x = 540$$
   $$330 + 2x = 540$$
   $$2x = 210$$
   $$x = 105°$$

3. **D** The volume of the container is V $= 5 \cdot 10 \cdot 12 = 600$ cubic feet. If it is filled to $\frac{2}{3}$ of capacity $600 \cdot \frac{2}{3} = 400$ cubic feet of water are in it. Once 50 more cubic feet are added, the total volume of water becomes 450 cubic feet. To find the percentage, divide 450 by 600:

   $$\frac{450}{600} = 0.75 = 75\%$$

4. **D**

   Set up a proportion to find the entire circumference C.

   $$\frac{3}{4} = \frac{12\pi}{C} \quad \text{Cross multiply.}$$
   $$3C = 48\pi$$
   $$C = 16\pi$$

   You can calculate the radius of the circle from its circumference.

   $C = 2\pi r = 16\pi$

   $2\pi r = 16\pi$ Divide both sides by 2π

   $r = 8$

   The original arc is $\frac{3}{4}$ of the circle. So the remaining angle, which is one of the angles of the triangle in the figure, is $\frac{1}{4}$ of 360, which is 90°.

   Use the Pythagorean theorem to find the distance between the endpoints of the arc:

   $8^2 + 8^2 = x^2$

   $128 = x^2$

   $x = 8\sqrt{2}$

   Notice that you can also use the ratios of the 45-45-90 triangle to solve this equation.

5. **D** Let the side of square A be 1. If it is increased by 100%, the side of square B will be 2. The area of B will be $2 \cdot 2 = 4$. One side of C becomes 4 since it is doubled from B. The area of C is $4 \cdot 4 = 16$.

Make sure you go back to the question, read the last sentence, and answer the correct question.

$$\frac{\text{Area } A}{(\text{Area } B + \text{Area } C)} = \frac{1}{(4 + 16)} = \frac{1}{20}.$$

6. **A**

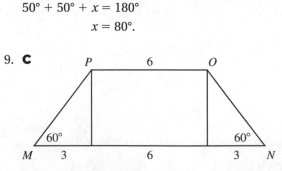

Since all angles are right angles, $a + b$ must be $14 + 22$ since they are parallel. Additionally, $c + d + e$ must equal 34.

The perimeter $= 34 + c + d + e + 22 + 14 + a + b = 34 + 34 + 36 + 36 = 140$

Alternatively, notice that the shape can be converted into a rectangle of sides 34 and 36. (Imagine that it is made up of thin wire.) The perimeter is $2 \cdot (34 + 36) = 140$.

7. **D** Draw the radius from M to K, which will be perpendicular to KT. Now you have right triangle MKT that has a hypotenuse of 17 and a leg of 15. Find MK (the radius) using the Pythagorean theorem:

$$15^2 + MK^2 = 17^2$$
$$MK^2 = 17^2 - 15^2$$
$$MK^2 = 64$$
$$MK = r = 8$$

8. **C** If the measure of arc CAB equals 100°, then angle CEB will equal 50°. The question also states that $FE = FC$, which makes CEF an isosceles triangle where angle FCE is also 50°. Use the interior angles of FEC (which equal 180°) to find x.

$$50° + 50° + x = 180°$$
$$x = 80°.$$

9. **C**

Draw the heights of the trapezoid from P and O. Notice that they create two 30-60-90 triangles with bases equal to 3.

The ratio of the height to the base of 3 can be obtained from the special triangle ratios ($x\sqrt{3}$). The height $= 3\sqrt{3}$.

Area $= \frac{1}{2}(6 + 12) \cdot 3\sqrt{3} = 27\sqrt{3}$

10. **B** Find the area of the quarter-circle and subtract the area of triangle ABD from it.

The area of the quarter-circle $= \frac{1}{4} \pi r^2 = \frac{1}{4} \pi 6^2 = 9\pi$.

The area of the triangle $= \frac{1}{2} \cdot 6 \cdot 6 = 18$.

The area of the shaded region is $9\pi - 18 = 9 \cdot 3.14 - 18 = 28.27 - 18 \approx 10$.

11. **E** Find the volume of the rectangular prism and divide by the volume of one cube:

$\frac{12 \cdot 8 \cdot 4}{2 \cdot 2 \cdot 2} = 6 \cdot 4 \cdot 2 = 48$

12. **C** Triangle ABC is a 45-45-90 triangle. So $\frac{AC}{BC} = \frac{\sqrt{2}}{1}$. Let BC be $2x$. AC can be calculated as $2x\sqrt{2}$.

The radius of the circle on $AC = x\sqrt{2}$, and the radius of the circle on $BC = x$.

$\frac{A_1}{A_2} = \frac{\pi \cdot r_1^2}{\pi \cdot r_2^2} = \frac{\pi \cdot (x\sqrt{2})^2}{\pi \cdot x^2} = \frac{2x^2}{x^2} = 2$

Note that the ratio of the areas of two circles is the square of the ratio of their radii,

$\left(\frac{\sqrt{2}}{1}\right)^2 = 2$.

13. **D**

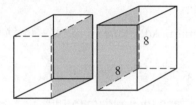

You do not need to find either the initial or the final surface area of the solids. The only addition to the surface area will be the shaded areas in the figure where the cut is made. You also do not need to know how far from the edge the cut is made. As long as it is parallel, the cut creates two additional surfaces that are squares.

So the additional area is $8 \cdot 8 \cdot 2 = 128$.

14. **D** The shortest distance on the surface of the cube from A to Z passes through the midpoint of the edge between them as shown in Figure 1. An easier way to see this is to redraw the surfaces the bug travels on a two-dimensional plane as in Figure 2.

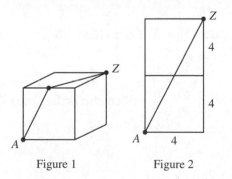

Figure 1   Figure 2

The volume of the cube is 64. The length of one edge is 4 cm since $4 \cdot 4 \cdot 4 = 64$.

In Figure 2, use the Pythagorean theorem to find the distance AZ:

$AZ^2 = 4^2 + 8^2 = 16 + 64 = 80$

$AZ = 4\sqrt{5}$

15. **D** Find one of the angles of a regular pentagon using

$$\frac{(n-2) \cdot 180}{n} = \frac{(5-2) \cdot 180}{5} = 108°$$

Let m∠ECD be x. △ECD is an isosceles triangle, since $ED = DC$. So m∠CED is also x.

In △ECD, $x + x + 108 = 180°$.

$$2x = 72$$
$$x = 36$$

$$△ABE \cong △CDE$$
$$m∠AED = 108°$$
$$36 + 36 + m∠BEC = 108°$$
$$m∠BEC = 36°$$

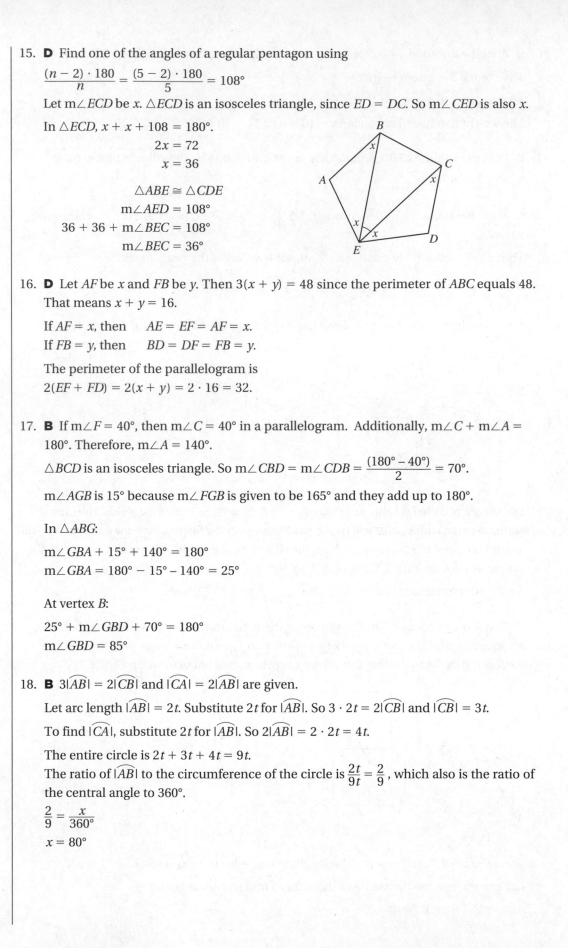

16. **D** Let AF be x and FB be y. Then $3(x + y) = 48$ since the perimeter of ABC equals 48. That means $x + y = 16$.

If $AF = x$, then $AE = EF = AF = x$.
If $FB = y$, then $BD = DF = FB = y$.

The perimeter of the parallelogram is
$2(EF + FD) = 2(x + y) = 2 \cdot 16 = 32$.

17. **B** If m∠F = 40°, then m∠C = 40° in a parallelogram. Additionally, m∠C + m∠A = 180°. Therefore, m∠A = 140°.

△BCD is an isosceles triangle. So m∠CBD = m∠CDB = $\frac{(180° - 40°)}{2} = 70°$.

m∠AGB is 15° because m∠FGB is given to be 165° and they add up to 180°.

In △ABG:

m∠GBA + 15° + 140° = 180°
m∠GBA = 180° − 15° − 140° = 25°

At vertex B:

25° + m∠GBD + 70° = 180°
m∠GBD = 85°

18. **B** $3|\overparen{AB}| = 2|\overparen{CB}|$ and $|\overparen{CA}| = 2|\overparen{AB}|$ are given.

Let arc length $|\overparen{AB}| = 2t$. Substitute $2t$ for $|\overparen{AB}|$. So $3 \cdot 2t = 2|\overparen{CB}|$ and $|\overparen{CB}| = 3t$.

To find $|\overparen{CA}|$, substitute $2t$ for $|\overparen{AB}|$. So $2|\overparen{AB}| = 2 \cdot 2t = 4t$.

The entire circle is $2t + 3t + 4t = 9t$.
The ratio of $|\overparen{AB}|$ to the circumference of the circle is $\frac{2t}{9t} = \frac{2}{9}$, which also is the ratio of the central angle to 360°.

$$\frac{2}{9} = \frac{x}{360°}$$
$$x = 80°$$

19. **E** Draw the altitudes AX and BY. Since $AD = 12$, use the ratios of the sides of a 30-60-90 triangle to find AX, which is equal to BY.

$$\frac{AX}{AD} = \frac{\sqrt{3}}{2}$$

$$\frac{AX}{12} = \frac{\sqrt{3}}{2}$$

$$AX = \frac{12 \cdot \sqrt{3}}{2} = 6\sqrt{3}$$

$$BY = 6\sqrt{3}$$

$$\frac{DX}{AD} = \frac{1}{2}$$

$$\frac{DX}{12} = \frac{1}{2}$$

$$DX = \frac{12}{2} = 6$$

YC is also $6\sqrt{3}$ since $\triangle BYC$ is an isosceles right triangle.

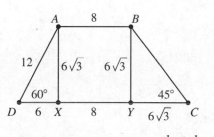

The area of the trapezoid is $\dfrac{b_1 + b_2}{2} \cdot h$.

$$\frac{8 + 14 + 6\sqrt{3}}{2} \cdot 6\sqrt{3} = \frac{22 + 6\sqrt{3}}{2} \cdot 6\sqrt{3} = 66\sqrt{3} + 54$$

20. **D** Divide the parallelogram into smaller regions parallel to the sides as shown.

If the area of $\triangle BCD = 12$, then the area of $BCDX = 24$. This means each of the small parallelograms has an area of 12. Count the number of parallelograms included in region $ABDEF$, which is 5. Area of $ABDEF = 12 \cdot 5 = 60$.

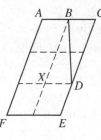

21. **E** Let the radius of one of the small circles be r. Then the radius of the large circle becomes $3r$.

The ratio of the total area of the small circles to the area of the large circle is

$$\frac{5\pi r^2}{\pi (3r)^2} = \frac{5r^2}{9r^2} = \frac{5}{9}$$

22. **D** First draw diagonal AC. It is equal to twice the radius of each circle. Use the 45-45-90 triangle ratios to find $AC = 2\sqrt{2}$.

If the radius of one quarter-circle is $\sqrt{2}$, its area is

$$A = \frac{1}{4}\pi(\sqrt{2})^2 = \frac{\pi}{2}$$

Since there are 2 quarter-circles, the total area of the circular regions is π.

The area of the square is $2 \cdot 2 = 4$, so the shaded area is $4 - \pi$.

23. **E** As you can see in the figure, only 8 disks could be cut from this sheet since the width of $3\pi \approx 9.42$ does not allow for a third row of disks.

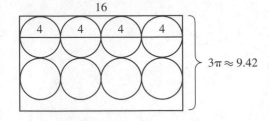

The area of the remaining sheet is $16 \cdot 3\pi - 8 \cdot \pi(2)^2$.

$48\pi - 32\pi = 16\pi$

24. **C** The cross section of the wood block looks like:

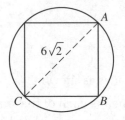

The largest cube that can be cut will have an edge of 6 ft. ($\triangle ABC$ is a 45-45-90 triangle.)

The question states that the cubes must be identical. You can cut only 2 of the $6 \times 6 \times 6$ cubes. Since the length of the block is only 16 ft, you cannot cut 3 cubes, which would be 18 ft long.

The volume of the 2 cubes is $2 \cdot 6^3 = 432$.

You may decide to cut 3 cubes that are $5 \times 5 \times 5$. In that case, the volume would be $3 \cdot 5^3 = 375$.

So the maximum total cube volume is 432 ft³.

Cubes that are $4 \times 4 \times 4$ would allow for 4 cubes to be cut, but the total volume would still be less than 432.

SECTION 5.3: COORDINATE GEOMETRY AND GRAPHS
Points on the Rectangular Coordinate System

- The **rectangular coordinate system** is used to locate and identify points on a plane relative to a fixed location (the **origin**.)

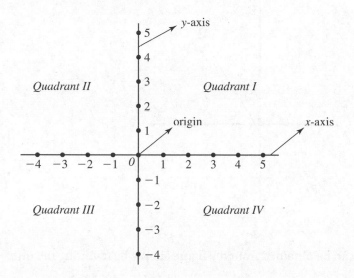

- Each point on the plane is identified by its orientation relative to the axes.
- Point M is 3 units above the x-axis.
  Point M is 2 units to the right of y-axis.
- The coordinates of point M are (2, 3).
- A pair of coordinates (x, y) is called an **ordered pair**.
- The coordinates of the origin are (0, 0).

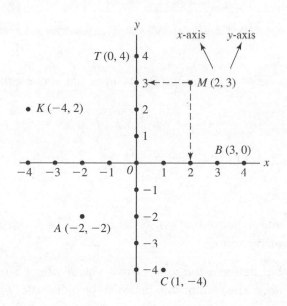

Distance Between Two Points

- The distance between two points (such as P and Q in the figure) can be found using the Pythagorean theorem. Notice that triangle PRQ is a right triangle.

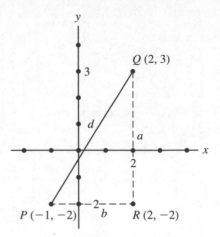

$$d^2 = a^2 + b^2$$

The lengths of a and b can be obtained from the figure simply by counting the units.

$a = 5$ and $b = 3$

$d^2 = 5^2 + 3^2 = 25 + 9 = 34$

$d = \sqrt{34}$

- A general formula to find the distance between two points $A(x_1, y_1)$ and $B(x_2, y_2)$ is
$d = \sqrt{(y_2 - y_1)^2 + (x_2 - x_1)^2}$

The distance between $P(-1, -2)$ and $Q(2, 3)$ is $d = \sqrt{(3 - (-2))^2 + (2 - (-1))^2} = \sqrt{5^2 + 3^2} = \sqrt{34}$

Midpoints

- The midpoint between two given points $A(x_1, y_1)$ and $B(x_2, y_2)$ can be found using the following formula:

$$M\left(\frac{x_1 + x_2}{2}, \frac{y_1 + y_2}{2}\right)$$

The midpoint between P and Q above is $\left(\frac{-1 + 2}{2}, \frac{-2 + 3}{2}\right) = \left(\frac{1}{2}, \frac{1}{2}\right)$.

Graphs

- Graphing an equation means marking all points that satisfy the equation. It is a visual way of showing a relationship between y and x.

  Example: Consider the relationship between the ages of two sisters where one is 3 years older than the other. The relationship between the older sister's age (y) and the younger sister's age (x) is simply y is three more than x. This relationship can be represented with a function or with a graph.

  As a function it is $y = x + 3$. If the age of the younger one is given, such as $x = 3$, you can easily figure out that the older one's age is 6 years ($y = 6$).

Before representing this relationship using a graph, take a look at a few points that satisfy the equation and set up a table.

x	y	
0	3	When the younger one was born, the older one was 3 years old. When $x = 0$, $y = 3$.
1	4	Plug $x = 1$ into $y = x + 3 = 1 + 3 = 4$.
3	6	Plug $x = 3$ into $y = x + 3 = 3 + 3 = 6$.
4	7	Plug $x = 4$ into $y = x + 3 = 4 + 3 = 7$.

This means the ordered pairs (0, 3), (1, 4), (3, 6), and (4, 7), as well as other points, are on your graph. Plot these points on the rectangular system and observe the nature of the relationship. This particular function is a linear function, therefore the graph will be a line.

The graphical representation of $y = x + 3$ is shown below.

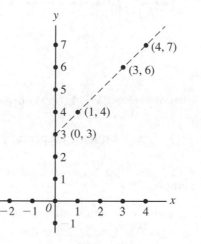

- A graph can be used to find particular points that satisfy a relationship. In some cases, only a graph is given to define a relationship.

Example: When the younger sister is 2 years old, $(x = 2)$, you can read the older sister's age from the graph. Find $x = 2$ on the x-axis. Move perpendicular to the x-axis (down or up) until you hit the graph. Read the y-value. The older sister is 5 years old.

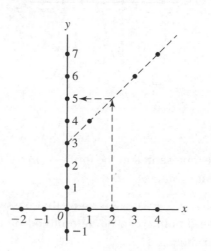

The same principle can be applied if only y is known. For example, if the older sister is 6.5 years old, find $y = 6.5$ on the y-axis. Move perpendicular to the y-axis (right or left) until you hit the graph. Read the x-value. The younger sister is 3.5 years old.

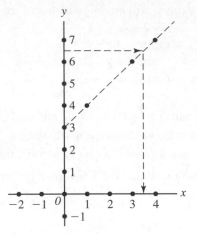

GRAPHS OF FUNCTIONS

- In a function in the form of $f(x) = 3x - 4$, $f(x)$ represents y, the output or the range of the function.
- Graphing a function $f(x) = 3x - 4$ means graphing $y = 3x - 4$ on the coordinate plane ($f(x) = y$).

Example: Sketch the graph of $f(x) = 3x - 4$.

First find two points that satisfy the equation. For example, plug in $x = 0$, $y = 3 \cdot 0 - 4 = -4$, and $x = 2$, $y = 3 \cdot 2 - 4 = 2$. So $(0, -4)$ and $(2, 2)$ are on the graph.

Second, locate these points on the coordinate axis.
Finally, draw a line through the points.

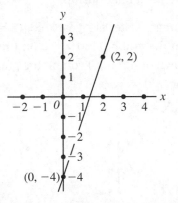

Lines

- Equations that can be represented in the form $y = mx + b$ are called **linear equations,** and their graphs are straight lines.

Example: The graph of $y = -2x + 3$ is a line that passes through $(-1, 5)$, $(0, 3)$, and $(1, 1)$. These points can be determined by creating a table of values as described above.

- Only one line can pass through any two given points. Since two points are enough to define a line, it is sufficient to find two points on the line to graph it.
- Mark (1, 1) and (−1, 5) on the coordinate system, and draw a line through these two points.
- As you can see, the line $y = -2x + 3$ also passes through (0, 3), (1.5, 0), and other points that can be found by looking at the graph.

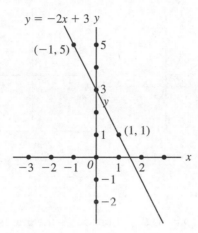

Slope

- Slope is defined as the ratio of the change in y to the change in x.

$$\text{Slope} = \frac{\text{Change in } y}{\text{Change in } x} = \frac{\text{Rise}}{\text{Run}} = \frac{y_2 - y_1}{x_2 - x_1}$$

To find the slope of line PQ, you can simply find a and b and divide.

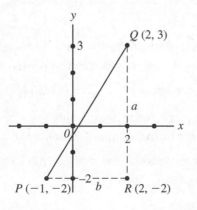

$$\text{Slope} = \frac{\text{Rise}}{\text{Run}} = \frac{a}{b} = \frac{5}{3}$$

You can also use the formula:

$$\text{Slope} = \frac{y_2 - y_1}{x_2 - x_1} = \frac{3 - (-2)}{2 - (-1)} = \frac{5}{3}$$

- In a linear equation in the form $y = mx = b$, m represents the slope.

   Example: In $y = -2x + 3$, the slope equals -2.

SLOPE

Slope is $\frac{\text{Rise}}{\text{Run}}$.

As you move to the right on a line from one point to the other, how far up you go versus how far right you go is the slope.

If the graph is going down, the slope is negative.

Positive slope

Negative slope

Zero slope
Example: $y = 3$

Undefined slope
Example: $x = 2$

Intercepts

- The y-intercept of a graph is the point where the graph crosses the y-axis.
   To find the y-intercept from the equation, set $x = 0$.

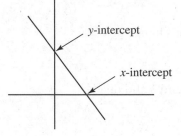

- If the equation is given in the form $y = mx + b$, b is always the y-intercept.
   Set $x = 0$ in $y = mx + b = m \cdot 0 + b = b$. The y-intercept is at $y = b$.

   Example: The y-intercept of $y = -42x + 17$ is 17.

- The x-intercept of a graph is the point where the graph crosses the x-axis.
   To find the x-intercept from the equation, set $y = 0$.

   Example: Find the y-intercept and x-intercept of $y = 7x - 10$.

   For the y-intercept, set $x = 0$. So $y = 7 \cdot 0 - 10 = -10$. The graph crosses the y-axis at $y = -10$.

   For the x-intercept, set $y = 0$. So $0 = 7 \cdot x - 10$, $x = \frac{10}{7}$. The graph crosses the x-axis at $x = \frac{10}{7}$.

WRITING THE EQUATION OF A LINE

- If two points on a line, such as (x_1, y_1) and (x_2, y_2), are given, use the following steps to find its equation.

    1. Find the slope using the slope formula: Slope $= m = \dfrac{y_2 - y_1}{x_2 - x_1}$.
    2. Use one of the points to write the equation: $y - y_1 = m \cdot (x - x_1)$.

    Example: Find the equation of the line that passes through $(3, 3)$ and $(2, -1)$.

    Slope $= m = \dfrac{y_2 - y_1}{x_2 - x_1} = \dfrac{-1 - 3}{2 - 3} = \dfrac{-4}{-1} = 4$

    By using $y - y_1 = m \cdot (x - x_1)$ and plugging in one of the points, you get $y - 3 = 4 \cdot (x - 3)$. This can be rearranged to become $y = 4x - 9$.

- If the slope of a line and one point on the line such as (x_1, y_1) are given, it is possible to write the equation of the line. Since you know the slope, simply use $y - y_1 = m \cdot (x - x_1)$.

    Example: Find the equation of the line that passes through $(3, 2)$ and has a slope of -2.

    Using $y - y_1 = m \cdot (x - x_1)$ and plugging in the point gives you $y - 2 = -2 \cdot (x - 3)$. It can also be rearranged to be $y = -2x + 8$.

PARALLEL AND PERPENDICULAR LINES

- Parallel lines have the same slope. If $\ell_1 \parallel \ell_2$, then $m_1 = m_2$.
- Perpendicular lines have negative reciprocal slopes. If $\ell_2 \perp \ell_3$, then $m_2 = -\dfrac{1}{m_3}$.

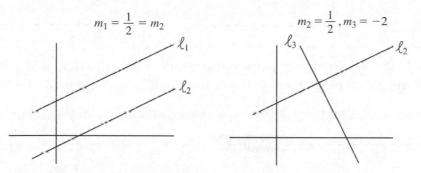

Quadratic Functions and Parabolas

- Graphs of quadratic equations are parabolas, as shown in the figure.

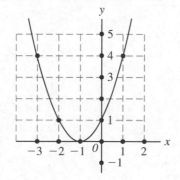

- To graph a quadratic equation, create a table of x-values and y-values to find a few points on the parabola.

Example: Graph $y = (x + 1)^2$

x	y
-2	1
-1	0
0	1
1	4

$y = (-2 + 1)^2 = (-1)^2 = 1$
$y = (-1 + 1)^2 = (0)^2 = 0$
$y = (0 + 1)^2 = (1)^2 = 1$
$y = (1 + 1)^2 = (2)^2 = 4$

Plot these points on the coordinate axis, and draw a parabola through them.

SAMPLE PROBLEMS

EXAMPLE 1

In the figure below, what is the area of $\triangle ABC$?

(A) 5
(B) 7.5
(C) 10
(D) 12.5
(E) 15

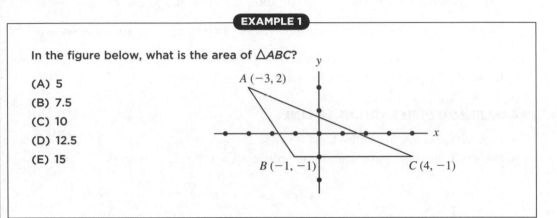

Base BC is 5 units long. The height is the vertical distance from BC to A (the difference in the y-coordinates). Therefore, the height is $2 - (-1) = 3$. The area of the triangle $= \frac{1}{2} 3 \cdot 5 = 7.5$. The answer is (B).

EXAMPLE 2

If point K $(n + 2, n - 3)$ is in Quadrant IV, how many integer values can n take?

(A) 1
(B) 2
(C) 3
(D) 4
(E) 5

If K is in Quadrant IV, then $x > 0$ and $y < 0$.
$$n + 2 > 0$$
$$n > -2$$
and
$$n - 3 < 0$$
$$n < 3$$
Since $-2 < n < 3$, n could be -1, 0, 1, or 2.
The answer is (D).

What is the *x*-intercept of the line that passes through (−1, −3) and (3, 5)?

(A) 2

(B) 1

(C) $\frac{1}{2}$

(D) $-\frac{1}{2}$

(E) −1

To find the equation of the line, find the slope first:

Slope $= m = \dfrac{y_2 - y_1}{x_2 - x_1} = \dfrac{5 - (-3)}{3 - (-1)} = \dfrac{8}{4} = 2$

Write the equation using $y - y_1 = m \cdot (x - x_1)$:

$y - 5 = 2 \cdot (x - 3)$ or $y = 2x - 1$

The *x*-intercept could be found by plugging in 0 for *y*:

$0 = 2x - 1$

$x = \dfrac{1}{2}$

The answer is (C).

SECTION 5.3—PRACTICE PROBLEMS

1. If point $B\,(m - 4, m)$ is in Quadrant II, what is the product of all possible integer values of *m*?

   (A) 2

   (B) 3

   (C) 5

   (D) 6

   (E) 8

2. In the rectangular coordinate system shown, $2 \cdot ON = 3 \cdot OM$. If the area of triangle *ONM* is 48, what is the *x*-coordinate of *M*?

   (A) 4

   (B) 6

   (C) 8

   (D) 10

   (E) 12

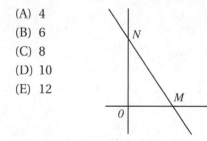

3. The equation of line *m* that passes through points $A\,(1, 2)$ and $B\,(2, -3)$ is $ay = bx = 14$. What is the value of *a*?

   (A) −5

   (B) $-\dfrac{3}{2}$

   (C) 2

   (D) $\dfrac{3}{3}$

   (E) 5

4. The midpoint between A (3, 2) and B (5, −6) is on the line given by the equation $kx + 3y − 6 = 0$. What is the value of k?

   (A) −3
   (B) −2
   (C) 1
   (D) 2
   (E) 3

5. Which of the following could be the value of n if the distance between points A $(3n, 5)$ and B $(5n, −3)$ is 10?

   (A) 2
   (B) 3
   (C) 4
   (D) 5
   (E) 6

6. The slope of the line that passes through $K(−3, −4)$ and $T(w, −2)$ is −1. What is the value of w?

   (A) −5
   (B) −3
   (C) −1
   (D) 1
   (E) 3

7. In the figure, $OA \perp AB$, m$\angle AOB = 60°$, and the coordinates of point B are (12, 0). What are the coordinates of point A?

   (A) $(4, 4\sqrt{3})$
   (B) $(3, 3\sqrt{3})$
   (C) $(3\sqrt{3}, 3)$
   (D) $(4\sqrt{3}, 4)$
   (E) $(2\sqrt{3}, 3\sqrt{3})$

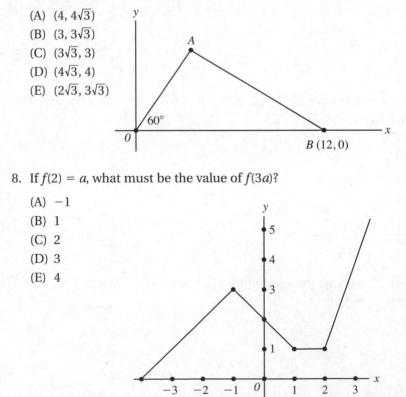

8. If $f(2) = a$, what must be the value of $f(3a)$?

   (A) −1
   (B) 1
   (C) 2
   (D) 3
   (E) 4

9. $ON = NM = MK$, and triangle MNP is an equilateral triangle. What is the x-coordinate of point T?

(A) $2\sqrt{3}$

(B) 4

(C) $4\sqrt{3}$

(D) 8

(E) $8\sqrt{3}$

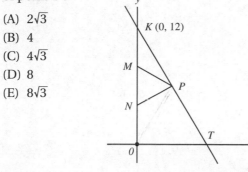

10. In the figure below, OT is tangent to the circle at point $T(12, 5)$. If the center of the circle is at $C(15, 0)$, what is the radius of the circle?

(A) $2\sqrt{14}$

(B) $4\sqrt{3}$

(C) 8

(D) $4\sqrt{7}$

(E) $8\sqrt{3}$

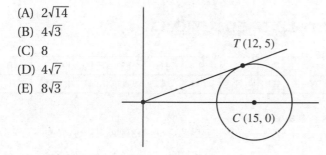

11. The vertices of parallelogram $EFGH$ are $E(-2, 0)$, $F(2, 3)$, $G(4, 1)$, and $D(k, t)$. Which one of the following is a possible value of $k + t$?

(A) -2

(B) -1

(C) 0

(D) 1

(E) 2

12. $ax + by = 42$ is a line that passes through points $P(1, 2)$ and $Q(-4, -3)$. What is $\frac{a}{b}$?

(A) $-\frac{4}{3}$

(B) -1

(C) $-\frac{3}{4}$

(D) $-\frac{1}{4}$

(E) $\frac{3}{4}$

13. The shaded area shown in the figure is equal to 9 square units. What is the value of a?

(A) 2

(B) 3

(C) 4

(D) 6

(E) 8

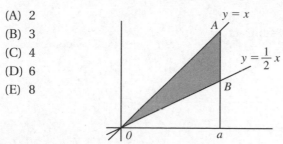

14. The equation of the line on the graph is $y = .{-}2x + 12$. If the perimeter of rectangle *OMNP* equals 14, what is the area?

(A) 8
(B) 10
(C) 12
(D) 14
(E) 16

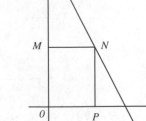

SECTION 5.3—SOLUTIONS

1	2	3	4	5	6	7	8	9	10	11	12	13	14
D	C	C	E	B	A	B	E	C	A	A	B	D	B

1. **D** If *B* is in Quadrant II, then $x < 0$ and $y > 0$.

$$m - 4 < 0$$
$$m < 4$$

and

$$m > 0$$

That means $0 < m < 4$.

Since *m* is a positive integer, *m* could be 1, 2, or 3.

$$1 \cdot 2 \cdot 3 = 6$$

2. **C** Let *OM* be $2x$. Then $2 \cdot ON = 3 \cdot 2x$ and $ON = 3x$.

The area of triangle *ONM* is

$$\frac{1}{2} \cdot 2x \cdot 3x = 48$$
$$3x^2 = 48$$
$$x^2 = 16$$
$$x = \pm 4 \qquad \text{Since } x \text{ is a length, it cannot equal } -4.$$

If $x = 4$, then $OM = 2x = 2 \cdot 4 = 8$.

3. **C** Plug $(1, 2)$ and $(2, -3)$ into $ay = bx + 14$:

$$2a = b + 14$$
$$-3a = 2b + 14$$

Multiply the first equation by 2, and subtract.

$$4a = 2b + 28$$
$$-\underline{-3a = 2b + 14}$$
$$7a = 14$$
$$a = 2$$

> **REMEMBER**
>
> If a line passes through a point, the coordinates of the point will satisfy the equation of the line.

4. **E** The midpoint can be found by:

$$M\left(\frac{x_1 + x_2}{2}, \frac{y_1 + y_2}{2}\right) = \left(\frac{3 + 5}{2}, \frac{2 - 6}{2}\right) = (4, -2)$$

If $(4, -2)$ is on the line $kx + 3y - 6 = 0$,

$$4k + 3(-2) - 6 = 0$$
$$4k - 12 = 0$$
$$4k = 12$$
$$k = 3$$

5. **B** Use the distance formula:

$$10 = \sqrt{(-3 - 5)^2 + (5n - 3n)^2} = \sqrt{(-8)^2 + (2n)^2} \qquad \text{Square both sides.}$$
$$100 = 64 + 4n^2 \qquad \text{Subtract 64.}$$
$$36 = 4n^2 \qquad \text{Divide by 4.}$$
$$n^2 = 9 \qquad \text{Take the square root.}$$
$$n = \pm 3$$

Only 3 is in the answer choices.

6. **A** Use the slope equation:

$$\text{Slope} = \frac{y_2 - y_1}{x_2 - x_1}$$

$$-1 = \frac{-2 - (-4)}{w - (-3)} = \frac{2}{w + 3} \qquad \text{Cross multiply.}$$

$$-w - 3 = 2$$
$$w = -5$$

7. **B**

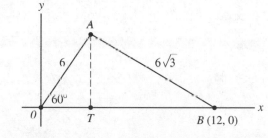

First use the 30-60-90 special triangle ratios for triangle OAB.

$$OB : OA : AB = 2x : x : x\sqrt{3}$$

$$\frac{12}{2x} = \frac{OA}{x}$$

$$OA = 6$$

$$\frac{12}{2x} = \frac{AB}{x\sqrt{3}}$$

$$AB = 6\sqrt{3}$$

Draw a perpendicular line from A to the x-axis to determine the x- and y-coordinates. Since T is a right angle, use the 30-60-90 triangle ratios in triangle OAT.

$$OA : OT : AT = 2x : x : x\sqrt{3}$$

$$\frac{6}{2x} = \frac{OT}{x}$$

$$OT = 3$$

$$\frac{6}{2x} = \frac{AT}{x\sqrt{3}}$$

$$AT = 3\sqrt{3}$$

The coordinates of point A are $(3, 3\sqrt{3})$.

8. **E**

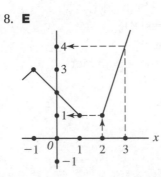

On the figure, draw a perpendicular line at 2 to the x-axis and read the y-coordinate on the graph. $f(2) = 1$

If $a = 1$, now you are looking for $f(3a) = f(3)$, which is equal to 4.

9. **C**

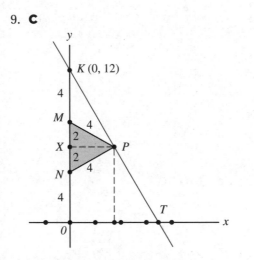

Since $OK = 12$, $ON = NM = MK = MP = NP = 4$.

In the equilateral triangle MNP, the height XP becomes $2\sqrt{3}$ from the 30-60-90 triangle.

Triangles KXP and KOT are similar triangles since XP is parallel to OT:

$$\frac{KX}{KO} = \frac{XP}{OT}$$

$$\frac{6}{12} = \frac{2\sqrt{3}}{OT}$$

$$OT = 4\sqrt{3}$$

10. **A** The length of OT can be found using the Pythagorean theorem:

$$OT = \sqrt{12^2 + 5^2} = \sqrt{169} = 13$$

If you draw the radius from C to T, CT is perpendicular to OT since T is the point of tangency. Use another Pythagorean theorem to find r:

$$r = \sqrt{15^2 - 13^2} = \sqrt{56} = 2\sqrt{14}$$

11. **A** Plot the points on the rectangular coordinate system as shown below. Point D could be at $(0, -2)$, so $k + t = -2$.

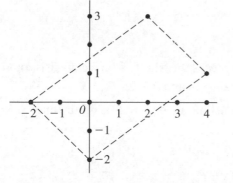

12. **B** If a point on a line is given, it satisfies the equation of the line. Plug $(1, 2)$ and $(-4, -3)$ into $ax + by = 42$ separately to get two separate equations.

$$a + 2b = 42 \quad \text{and} \quad -4a - 3b = 42$$

Since they are both equal to 42, you can set them equal to each other:

$$a + 2b = -4a - 3b$$
$$5a = -5b$$
$$\frac{a}{b} = -1$$

Alternatively, rearrange $ax + by = 42$ into slope-intercept form:

$$y = -\frac{a}{b}x + \frac{42}{b}$$

This means $-\frac{a}{b}$ is the slope. You can find the slope of the line using two points:

$$m = \frac{-3 - 2}{-4 - 1} = \frac{5}{5} = -\frac{a}{b}$$

We are looking for $\frac{a}{b} = -1$.

13. **D** The area of a triangle is $\frac{1}{2} \cdot$ base $\cdot$ height. In the figure, the base triangle AOB is AB and the height is equal to a.

To find the base (distance AB), plug a into the given line equations to find their y-coordinates:

$$y_B = \frac{1}{2}a \text{ and } y_A = a$$

The difference between them is the base.

$$\text{Base} = a - \frac{1}{2}a = \frac{1}{2}a$$

The area is $\frac{1}{2} \cdot \frac{1}{2} a \cdot a = 9$.

$a^2 = 36$

$a = 6$

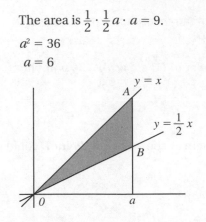

14. **B** Let the coordinates of point N be (a, b). That means the width of the rectangle is a and the length is b. If the perimeter is 34, then $2a + 2b = 14$ and $a + b = 7$.

You can also plug (a, b) into the equation of the line since point N is on the line:

$b = -2a + 12$

$b + 2a = 12$

Solve the two equations simultaneously:

$\begin{array}{r} 2a + b = 12 \\ - \quad a + b = 7 \\ \hline a = 5 \end{array}$

Since $a = 5$, $5 + b = 7$ so $b = 2$.

Area $= 2 \cdot 5 = 10$

SECTION 5.4: DATA INTERPRETATION

Some of the questions on the GMAT require you to analyze certain types of graphs and tables. Examples of some of the most common question types are given below.

Table

Energy Bars	Calories	Protein (g)	Total Fat (g)
A	105	14	7
B	220	12	9
C	104	4	13
D	202	10	4.5
E	240	20	11

The table above shows select nutritional facts for 5 energy bars. Answer the following three questions based on the data given.

Question 1: Which of the energy bars has the highest protein per total fat?

      (A) *A*   (B) *B*   (C) *C*   (D) *D*   (E) *E*

This problem is a perfect case to use the process of elimination. The ratio of protein to fat for bar A is $14 : 7$, which is exactly 2. Compare all other ratios to 2 and eliminate the ones that are less than 2. For bar B, the ratio is $\frac{12}{9}$ which is less than 2; eliminate B. The ratio for bar C is $\frac{4}{13}$ which is less than 1; eliminate C. The ratio for bar D is $\frac{10}{4.5}$ which is slightly greater than 2, since $\frac{10}{5}$ is exactly 2; eliminate A. The ratio for bar E is $\frac{20}{11}$ which is slightly less than 2; eliminate E.

The answer is (D).

Question 2: The average protein content of all five bars is what percent greater than that of bar D?
  (A) 120%   (B) 100%   (C) 20%   (D) 12%   (E) 1.2%

The average protein content of all five bars can be found by adding their protein contents and dividing by 5: $\frac{(14 + 12 + 4 + 10 + 20)}{5} = \frac{60}{5} = 12$. The protein content of bar D is 10 g. Now the question reduces to 12 is what percent greater than 10?

Percent difference $= \frac{12 - 10}{10} \cdot 100\% = 20\%$

The answer is (C).

Question 3: Which of the energy bars has the highest amount of protein per calorie?
  (A) A (B) B (C) C (D) D (E) E

Once again, use elimination to compare the bars among themselves. In doing so, you will not need to calculate the exact ratios but only rough estimates. For example, bar A offers 14 g of protein per 105 calories. That is roughly $\frac{14}{105} \approx \frac{14}{100} = 0.14$.

Now see if you can eliminate any answer choices by comparing them to 0.14. The ratio for bar B is $\frac{12}{220}$, which is roughly equal to $\frac{12}{200} = \frac{6}{100} = 0.06$. All you need to do is a rough mental calculation to see if the ratio is even close to 0.14. If it is significantly lower, you can eliminate it. If it could be close, you should calculate the exact numbers. In this case, the percentage for B is very low compared with A; eliminate B. The ratio for bar C is $\frac{4}{104} \approx \frac{4}{100}$ which is roughly 0.04; eliminate C. The ratio for bar D is $\frac{10}{202} \approx \frac{10}{200} = \frac{5}{100}$, which is roughly 0.05; eliminate D. The ratio for bar E is $\frac{20}{240}$. Compare this ratio to $\frac{20}{200} = 0.10$. The ratio $\frac{20}{240}$ is less because it has a larger denominator. Therefore $\frac{20}{240} < 0.10$; eliminate E.

The answer is (A).

Bottled Tea	Grams of Sugar per Serving	Number of Servings
A	140	2.0
B	150	3.0
C	90	2.5
D	150	1.5
E	220	2.0

The table above shows the sugar content per serving of various bottled teas and the number of servings included in one bottle. Use the data to answer the following three questions.

Question 1:   What is the average sugar content per serving?

    (A) 140   (B) 150   (C) 300   (D) 324   (E) 750

Average grams of sugar per serving can be found by simply taking the average of the second column:

Average = $\dfrac{140 + 150 + 90 + 150 + 220}{5} = \dfrac{750}{5} = 150$

The answer is (B).

Question 2:   What is the average sugar content per bottle?

    (A) 140   (B) 150   (C) 300   (D) 324   (E) 750

The average grams of sugar per bottle can be found by averaging the sugar content per bottle. To find the amount of sugar in one bottle, multiply the sugar content per serving by the number of servings. For example, bottle A has $140 \cdot 2 = 280$ grams per bottle. Find the amount in each bottle and average as follows:

$$\text{Average} = \frac{(140 \cdot 2) + (150 \cdot 3) + (90 \cdot 2.5) + (150 \cdot 1.5) + (220 \cdot 2)}{5}$$

$$= \frac{(280 + 450 + 225 + 225 + 440)}{5} = \frac{1620}{5} = 324$$

The answer is (D).

Question 3:   Which tea has the highest amount of sugar per bottle?

    (A) A (B) B (C) C (D) D (E) E

Bottle A has $(140 \cdot 2) = 280$ grams of sugar per bottle.
Bottle B has $(150 \cdot 3) = 450$ grams of sugar per bottle.
Bottle C has $(90 \cdot 2.5) = 225$ grams of sugar per bottle.
Bottle D has $(150 \cdot 1.5) = 225$ grams of sugar per bottle.
Bottle E has $(220 \cdot 2) = 440$ grams of sugar per bottle.

The answer is (B).

Circle Graphs

The circle graph shows the percentage of each ingredient by weight included in Joy's Trail Mix. Answer the following questions based on the graph.

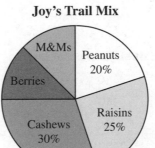

Joy's Trail Mix

Question 1: If Joy's Trail Mix has equal amounts of berries and M&Ms by weight, what is the ratio of the weight of M&Ms to the weight of raisins?

    (A) 1 : 4   (B) 1 : 3   (C) 1 : 2   (D) 2 : 3   (E) 3 : 4

The percentages of all ingredients add up to 100%. Add all given percentages: 20% + 30% + 25% = 75% and subtract from 100%. So 100% − 75% = 25%. Since M&Ms and berries have equal weights, then each of them make up $\frac{25\%}{2}$ = 12.5% of the total. The ratio of the weight of M&Ms to the weight of raisins is 12.5% : 25% = 1 : 2

The answer is (C).

Question 2: What is the degree measure of the central angle of the slice that represents cashews?

    (A) 30°   (B) 96°   (C) 100°   (D) 108°   (E) 112°

A circle contains 360°. Since cashews are 30% of the mixture, the cashew slice has a central angle of 30% · 360° = 0.3 · 360° = 108°.

You can also set up a proportion to find the angle:

$$\frac{30}{100} = \frac{x}{360}$$
$$x = 108°$$

The answer is (D).

Question 3: If the entire mixture weighs 40 ounces, how many ounces of nuts are used in the mixture?

    (A) 8   (B) 12   (C) 20   (D) 25   (E) 50

There are two types of nuts, peanuts (20%) and cashews (30%). The total percentage of nuts is 50%. Therefore, the weight of the nuts can be found by multiplying 50% by 40 ounces.

50% · 40 = 20 ounces

The answer is (C).

Question 4: Sanjay buys 64 ounces of Joy's Trail Mix. Since he dislikes berries, he picks them out. Which of the following is closest to the new percentage of M&Ms in Sanjay's trail mix?

    (A) 25%   (B) 22%   (C) 16%   (D) 14%   (E) 8%

Berries are 12.5% of the mixture initially. That means the total amount of berries equals 12.5% · 64 = 8 ounces and the total amount of M&Ms equals 12.5% · 64 = 8 ounces. After he picks out the berries, the new mixture weighs 64 − 8 = 54 ounces. The percentage of M&Ms can be found using a proportion: $\frac{8}{54} = \frac{x}{100}$.

Cross multiply 800 = 54x, So $x = \frac{800}{54}$. Since the question is asking for the approximate percentage, $\frac{800}{54} \approx \frac{800}{50} = 16$. Since we are actually dividing 800 by 54, which is larger than 50, our answer needs to be slightly smaller than 16. The closest answer choice is 14%.

The answer is (D).

Bar Graphs

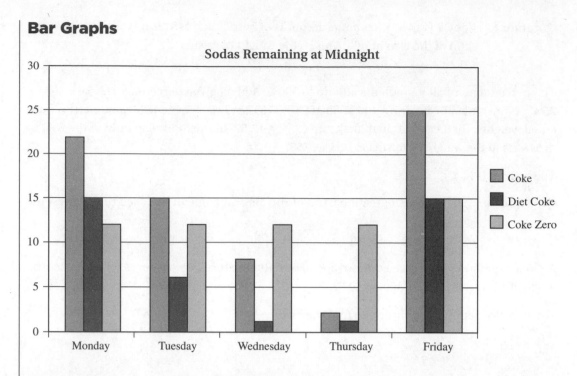

Sodas Remaining at Midnight

The bar graph above shows the number of cans of soda remaining in a vending machine at midnight after each workday. The cans are not replenished each night. Answer the following questions based on the chart.

Question 1: What is the average number of cans of Diet Coke left in the vending machine during the workweek?

(A) 1   (B) 5.6   (C) 6   (D) 6.7   (E) 7.6

The number of cans of Diet Coke left in the vending machine each night is 15, 6, 1, 1, 15. To find the average, add all the numbers and divide by 5.

$$\text{Average} = \frac{(15 + 6 + 1 + 1 + 15)}{5} = 7.6$$

The answer is (E).

Question 2: How many cans of Coke were sold on Tuesday?

(A) 7   (B) 6   (C) 5   (D) 4   (E) 3

Since the number of cans of Coke at midnight on Monday was 22 and at midnight on Tuesday was 15, $22 - 15 = 7$ cans of Coke were sold on Tuesday.

The answer is (A).

Question 3: What day did Diet Coke have its lowest sales?

(A) Monday   (B) Tuesday   (C) Wednesday   (D) Thursday   (E) Friday

On Thursday, the number of Diet Cokes did not change, so zero were sold.

The answer is (D).

Cumulative Graphs

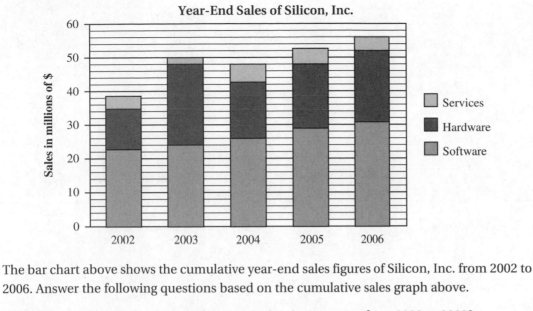

The bar chart above shows the cumulative year-end sales figures of Silicon, Inc. from 2002 to 2006. Answer the following questions based on the cumulative sales graph above.

Question 1:   What is the percent decrease in Services revenue from 2002 to 2003?

   (A) 100%   (B) 75%   (C) 50%   (D)25%   (E) 10%

Look at the height of the green sections. The services sales in 2002 was $4 million and in 2003 was $2 million, therefore the percent change is:

$$\% \text{ change} = \frac{\text{New} - \text{Original}}{\text{Original}} \cdot 100\% = \frac{2-4}{4} \cdot 100\% = -50\% \rightarrow 50\% \text{ decrease}$$

The answer is (C).

Question 2:   In which year was the percent increase in the hardware revenue the largest?

   (A) 2002   (B) 2003   (C) 2004   (D) 2005   (E) 2006

Hardware revenues from 2002 to 2006 are 12, 23, 17, 19, and 21. The largest increase is from 12 in 2002 to 23 in 2003, which also is the highest percentage increase in this case. It is an increase of more than 100%.

The answer is (D).

Question 3:   How much did total revenue decline in 2004?

   (A) $2 million   (B) $5 million   (C) $7 million   (D) $10 million   (E) $12 million

The total revenue in 2003 was $50 million and in 2004 was $48 million. The decline was $50 - 48 = 2$.

The answer is (A).

Question 4:   What fraction of the overall revenues in 2006 came from hardware sales?

   (A) $\frac{1}{4}$ (B) $\frac{1}{3}$ (C) $\frac{3}{8}$ (D) $\frac{1}{2}$ (E) $\frac{5}{8}$

In 2006, total sales were $56 million and hardware sales were $21 million. The ratio is $\frac{21}{56}$, which simplifies to $\frac{3}{8}$.

The answer is (C).

Data Sufficiency

6

→ **6.1 FORMAT**

→ **6.2 TWO TYPES OF QUESTIONS**

→ **6.3 INTERPRETING THE STEM**

→ **6.4 CHALLENGING THE STATEMENTS**

SECTION 6.1: FORMAT

If this is the first time you are taking the GMAT, data sufficiency questions are probably new to you. As the name implies, a data sufficiency problem asks you to determine if the given information is sufficient to find an answer. Approximately 30–40% of the questions in the math section of GMAT will be in this format. Data sufficiency questions do not actually require you to find the particular answer. They measure your ability to analyze a question, identify irrelevant information, and be able to decide which pieces of information are in fact sufficient to answer that question.

All data sufficiency questions have the same format. The first part (the stem) contains the question and in some cases some initial information. If information is provided in the stem, it is definite. This piece is not in question.

After the stem, two statements, statement (1) and statement (2), are provided. They contain two separate pieces of information that may or may not be relevant/sufficient to answer the question. Your job is to decide if you can answer the question using the first piece of information only, using the second piece of information only, or using both of them together. In some cases, each statement alone will be sufficient. In some, both statements will be needed. In other cases, neither statement will be sufficient, even when taken together.

> **REMEMBER**
>
> Do just enough work to determine if you *can* answer the question.

Take a look at the example below:

The stem

Kyle opened a new savings account 7 years ago and hasn't made any transactions since.

How much money does Kyle have today?

The stem provides two pieces of information. The time period is 7 years, and Kyle made no further transactions.

You must decide if you have enough information to calculate the amount of money today.

The statements

(1) The interest rate is 2.5%.

(2) Kyle deposited $2500.

First analyze statement (1) and statement (2) separately. Analyze them together **only if** neither of them is sufficient alone.

It cannot be stressed enough that you do not need to calculate an actual value or decide if the answer is yes or no to answer a data sufficiency question. You simply need to discover if you have enough information. Almost everyone's initial reaction is to try to find an exact value to make sure, but doing that could be a big time trap for you. Since test preparers know people have this tendency, some data sufficiency questions are designed to tempt you to spend time solving for an answer.

The answer choices will always be the same. Memorize them before you start practicing your data sufficiency problems to make sure you are very comfortable examining them on the test day. We have listed the official answer choice language below.

(A) Statement (1) ALONE is sufficient, but statement (2) alone is not sufficient.
(B) Statement (2) ALONE is sufficient, but statement (1) alone is not sufficient.
(C) BOTH statements TOGETHER are sufficient, but NEITHER statement ALONE is sufficient.
(D) EACH statement ALONE is sufficient.
(E) Statements (1) and (2) TOGETHER are NOT sufficient.

Step-by-Step

Use a methodical approach to review the statements presented.

STEP 1 Read the question very carefully and rephrase/simplify it if necessary (more on that later). Think about what additional information you would need to answer the question.

STEP 2 Consider only statement (1) with the stem.

If statement (1) alone is <u>sufficient</u>, then the only possible answer choices are <u>A</u> or <u>D</u>. You can cross out B, C, and E on your scratch pad.

If the first statement (1) alone is <u>not sufficient</u>, the only possible answer choices are <u>B, C, and E</u>. You have eliminated A and D.

} A, D

or

} B, C, E

MEMORIZE

1 alone	A
2 alone	B
Together	C
Each alone	D
Neither	E

STEP 3 Consider only statement (2) with the stem, disregarding the first statement. Avoid the common mistake of thinking about the first statement as a given when considering the second statement.

STEP 4 If statements (1) and (2) are not sufficient separately, then and only then consider them together. Combine the stem with the two additional pieces of information given and ask the sufficiency question again.

The matrix shows the possible choices for answering the sufficiency questions. Following the steps above, steps 2 and 3 ask you to consider statement (1) alone first and statement (2) alone next (the lighter region). Only if the answers are no–no would you end up in the darker region (step 4) and consider them together.

As you can see, if (1) alone is sufficient, the only answer choices are D and A (first column).

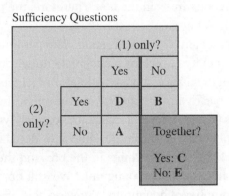

Sufficiency Questions

It is a very good idea to use the elimination technique each time. It removes the definitively wrong answers, thereby saving you time and easily avoidable mistakes.

SECTION 6.2: TWO TYPES OF QUESTIONS

There are two main types of data sufficiency problems, those asking for a specific value and yes/no questions.

VALUE QUESTIONS

Value questions ask for one unique quantity. Some examples of value questions include:

- How old was Ali 2 years ago?
- What is the value of $2x + 3y$?
- By what percent did Dana's salary increase?

Only one value can answer the question. The information is considered insufficient if it does not allow for a value to be calculated or if it allows for multiple values to be calculated.

IMPORTANT!

In value questions, the information is sufficient only if it is possible to determine a <u>unique</u> quantity for the question.

Example:

Tricia brings a box of doughnuts to a party. If a doughnut is picked at random, what is the probability that it will be a chocolate doughnut?

(1) There are 16 chocolate doughnuts and 12 vanilla doughnuts in the box.

(2) There are 14 people at the party, and each of them ate a chocolate doughnut.

STEP 1 We only know that there is a box of doughnuts. We would either need the number of each kind included in the box or the ratio of the number of each kind of doughnuts.

STEP 2 Consider only statement (1) with the stem.

What is the probability of picking a chocolate doughnut if there are 16 chocolate doughnuts and 12 vanilla doughnuts in the box?

Statement (1) alone is <u>not sufficient</u> since we cannot assume that there are only chocolate and vanilla doughnuts in the box. Therefore, the possible answer choices are <u>B, C, and E</u>. Eliminate A and D.

STEP 3 Consider only statement (2) with the stem, disregarding the first statement.

What is the probability of picking a chocolate doughnut if there are 14 people at the party and each of them ate a chocolate doughnut (since we are ignoring statement 1)?

Statement (2) is clearly not enough to answer the probability question. Eliminate B.

STEP 4 Since statements (1) and (2) are not sufficient separately, we need consider them together.

What is the probability of picking a chocolate doughnut if there are 16 chocolate doughnuts and 12 vanilla doughnuts in the box and there are 14 people at the party, each of whom ate a chocolate doughnut? We still do not have enough information about the total number of doughnuts. Knowing how many doughnuts were eaten and the number of available doughnuts does not help us calculate the probability of picking a chocolate doughnut in the first place.

No value can be calculated even when we use the two statements together.

The answer is (E).

Example:

What is the value of n?

(1) $\frac{182}{n}$ is a prime number.

(2) n is an even number.

STEP 1 No initial information is provided in this stem.

STEP 2 Consider only statement (1) with the stem.

What is the value of n if $\frac{182}{n}$ is a prime number?

Find the prime factorization of 182 first. $182 = 2 \cdot 7 \cdot 13$. So n could be 14, 26, and 91 since $\frac{182}{14} = 13$, and $\frac{182}{26} = 7$, and $\frac{182}{91} = 2$. There are 3 possible values for n, therefore the first statement (1) alone is <u>not sufficient</u>. The possible answer choices are <u>B, C, and E</u>. Eliminate A and D.

STEP 3 Consider only statement (2) with the stem, disregarding the first statement.

What is the value of n if n is an even number? Statement 2 is clearly not enough information to find out about n. Eliminate B.

STEP 4 Since statements (1) and (2) are not sufficient separately, we need consider them together.

What is the value of n if $\frac{182}{n}$ is a prime number and n is an even number?

We still do not have enough information to find a unique value for n since two of the three possible n values we identified in Step 2 are even.

No unique value for n can be calculated even when we use the two statements together: The answer is (E).

YES/NO QUESTIONS

Yes/No questions ask if you can give a definitive yes or a definite no answer. Examples include:

- Is $m > n$?
- Is x even?
- Does Sue make more money per month than Greg?

Sufficient means either an "absolute yes" or an "absolute no." **Insufficient** means sometimes yes and sometimes no.

This question type is potentially tricky for some students since some tend to confuse having enough information to say no with not having enough information.

The following example is a straightforward question.

Is Carl's rent more than $750?

(1) Carl's rent is less than Sue's rent.
(2) Sue pays $650 for rent.

STEP 1 This is a yes/no question. We are only interested in figuring out if Carl's rent is more or less than $750.

STEP 2 Consider only statement (1) with the stem.

If Carl's rent is less than Sue's, is he paying more than $750 for rent? We do not have enough information to answer that question. Statement (1) alone is not sufficient to answer yes or no since we don't know how much Sue pays. Therefore, the possible answer choices are B, C, and E. Eliminate A and D.

STEP 3 Consider only statement (2) with the stem, disregarding the first statement.

If Sue's rent is $650, is Carl paying more than $750 for rent? We do not have enough information to answer that question. Statement (2) alone is also not sufficient since we are using only the information provided in statement (2) without statement (1). Eliminate B.

STEP 4 Since statements (1) and (2) are not sufficient separately, we need to consider them together.

Here's the new question. If Carl's rent is less than Sue's and Sue pays $650 for rent, is Carl paying more than $750 for rent? Now we have enough information to answer the question, which means the answer choice is C. BOTH statements TOGETHER are sufficient, but NEITHER statement ALONE is sufficient. That Carl doesn't pay more than $750 does not mean your answer choice should be E.

The answer is (C).

Example:

Is x greater than y?

(1) $x = 2 - \left(\frac{63}{67} + \frac{77}{79} \right) = y$

(2) $x = 2 - \left(\left(\frac{43}{48} - \frac{101}{100} \right) \cdot y \right)^2$

> **IMPORTANT!**
> In yes/no questions, "sufficient" means either an absolute yes or an absolute no.

> **IMPORTANT!**
> In yes/no questions, if you can answer the question either way, you have a solution. "No" does not mean there's not enough information.

STEP 1 This is a yes/no question. We are only interested in figuring out if x is greater than y.

STEP 2 Consider only statement (1) with the stem.

Is x greater than y if $x = 2 - \left(\frac{63}{67} + \frac{77}{79}\right) + y$?

Take a look at the $2 - \left(\frac{63}{67} + \frac{77}{79}\right)$ piece first. We do not need to know the exact value of that number, but we do need to know if it is positive or negative. Since their numerators are less than their denominators, $\frac{63}{67}$ and $\frac{77}{79}$ are both less than 1. Therefore their sum will be less than 2 and 2 minus a number less than 2 is a positive number. The question reduces to asking if x is greater than y if x equals a positive number $+ y$? We have enough information to answer the question.

So statement (1) alone is sufficient. The only possible answer choices are A or D. You can eliminate B, C, and E.

STEP 3 Consider only statement (2) with the stem, disregarding the first statement.

Is x greater than y if $x = \left(\left(\frac{43}{48} - \frac{101}{100}\right) \cdot y\right)^2$?

You can rewrite the second piece as $\left(\frac{43}{48} - \frac{101}{100}\right)^2 \cdot y^2$. Since we cannot determine if y is positive or negative, the coefficient $\left(\frac{43}{48} - \frac{101}{100}\right)^2$ does not really matter. We cannot answer the question considering statement (2) alone.

STEP 4 Since statement (1) alone is sufficient, we do not need to consider them together. Statement (1) ALONE is sufficient, but statement (2) alone is not sufficient.

The answer is (A).

SECTION 6.3: INTERPRETING THE STEM

Rephrasing or writing down your interpretation of what the question is really asking is always useful and time efficient. If the stem contains an expression, simplify it as much as possible. If the stem contains a statement, see if you can rephrase it to clarify the question. Consider the following examples:

Is $\dfrac{100 + 2n}{n}$ an integer?	Simplify as $\dfrac{100}{n} + \dfrac{2n}{n} = \dfrac{100}{n} + 2$.
$\dfrac{3x^2 + 6xy + 3y^2}{x + y}$ might be given.	Simplify as $\dfrac{3(x^2 + 2xy + y^2)}{x + y} = \dfrac{3(x + y)^2}{x + y} = 3(x + y)$ where $x + y \neq 0$
Is $n^2 - n > 0$?	Simplify as $n(n - 1) > 0$. In other words, is the product of n and $(n - 1)$ positive?
Is k^2 odd?	k^2 can be odd only if k is odd. So the question is is k odd?
$(n - 1)$ is an even number.	n is an odd number.
Sue's rent is not more than \$750.	Sue's rent is less than or equal to \$750.
When n is divided by k, the remainder is zero.	n is divisible by k.
$(1 - x)(1 + y) > 0$	Either both $(1 - x)$ and $(1 + y)$ are both positive or both negative since their product is positive.

SECTION 6.4: CHALLENGING THE STATEMENTS

In certain data sufficiency questions, you can test a range of numbers to see if the statements are sufficient or not. When testing numbers, it is easier (and definitive) to prove that the statement is insufficient than to prove it is sufficient.

For example, consider the following question:

Is $(x - 1)^2 < 3$?

(1) $x < 5$
(2) $x > 2$

> **NUMBERS TO TRY**
>
> Positive
> Negative
> $0 < x < 1$
> $-1 < x < 1$
> -1, 0, and 1
> Fractions

If you've decided to test some numbers, make sure you try exceptions (within the parameters) for each statement to show that the statements are insufficient. If you cannot get contradicting results and for every exception you get a consistent answer (either always yes or always no), you must accept that it is sufficient. The exceptions are numbers such as negative numbers, numbers between -1 and 0, numbers between 0 and 1, other fractions, and the numbers -1, 1, and 0.

For statement (1), try 2 and -2.
$$(2 - 1)^2 < 3$$
$$1 < 3$$

So 2 gives a true statement. However, it does not prove that $x < 5$ would work for any real number. Try -2.
$$(-2 - 1)^2 < 3$$
$$9 < 3$$

So -2 gives a false statement. Since you have two contradicting answers, you can conclude that statement (1) is not sufficient.

For statement (2), try 3.
$$(3 - 1)^2 < 3$$
$$4 < 3$$

So 3 gives a false statement. Try a larger number such as 5.
$$(5 - 1)^2 < 3$$
$$16 < 3$$

So 5 also gives a false statement. They agree so far. Since the stem does not indicate that x is an integer, try 2.2 since you know that 2 gives you a true answer.
$$(2.2 - 1)^2 < 3$$
$$1.2^2 < 3$$

So 2.2 gives a true statement. You have contradicting answers again. Statement (2) also is not sufficient alone.

Consider statements (1) and (2) together. $x < 5$ and $x > 2$ means $2 < x < 5$. Is $(x - 1)^2 < 3$? In this case, you can try 2.2 and 4. These numbers will give you contradicting answers again. The answer is (E).

In some cases, all of the exceptions will give you consistent answers. Then you can conclude that the statement is sufficient.

Final Notes

In data sufficiency problems:

- All numbers are real numbers.
- Figures conform to the stem but not necessarily to the statements.
- All lines are straight.
- All angle measurements are positive.
- All figures lie in the same plane.

> **IMPORTANT!**
>
> If you succumb to the temptation to solve for a value, know that statements (1) and (2) never contradict each other. Therefore, if they are both sufficient, they should individually lead you to the same result.

> ### FIGURES NOT DRAWN TO SCALE
>
> Do not guesstimate based on the dimensions of the illustrations given in geometry questions. Figures are generally not drawn to scale. Redraw them on your scratch pad based on given information.

PRACTICE PROBLEMS

1. What is the product of a and b?
   (1) a and b are consecutive integers.
   (2) $a^2 + b^2 = 613$

2. PR and RP are two different two-digit numbers. What is the value of P?
   (1) $PR + RP = 132$
   (2) $P > R$

3. When k is divided by t, the quotient is 6 and the remainder is 4. What is value of k?
   (1) $t = 4$
   (2) $k = 2t + 20$

4. k and t are positive integers where
   $k + \frac{t}{4} = 3.75$. What is the sum of k and t?
   (1) k and t are both less than 5.
   (2) t is an odd integer.

5. x, y, and z are nonzero real numbers. $x \cdot y < 0$. Is $\frac{xz}{y}$ negative?
   (1) $x^2 z < 0$
   (2) $x \cdot z < 0$

6. P, Q, and R are points on the same line. $PR = 6$. What is the distance between P and Q?

    (1) $RQ = 2$
    (2) Q is between P and R.

7. How old is Clive today?

    (1) Deniz, Kaya, and Clive's ages are proportional to 2, 3, and 5 today.
    (2) Deniz is 5 years younger than Kaya.

8. In a certain test, a student makes three quarters for each correct answer and loses two quarters for each wrong answer. No loss or gain is recorded for unanswered questions. How much money did the student make or lose on this test?

    (1) The test had 60 questions.
    (2) The student had 5 more correct answers than wrong answers.

9. What is the smallest number in a set of 7 numbers?

    (1) The sum of the 7 numbers is 126.
    (2) All numbers in the set are even.

10.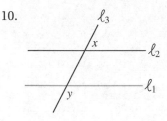

    What is the value of y?

    (1) ℓ_1 and ℓ_2 are parallel.
    (2) $y = x + 40$

11. A, B, and C are positive integers. A is what % of C?

    (1) A is 20% of B.
    (2) B is 30% of C.

12. There are a total of 150 doctors and nurses in a hospital. How many female doctors are there?

    (1) The ratio of nurses to doctors is $4 : 1$.
    (2) The ratio of female staff to male staff $1 : 2$.

13. What is the ratio of $\frac{z}{x}$ if $xy = 6 - \frac{x}{z}$?

    (1) $x \cdot y \cdot z = 20$
    (2) $y \cdot z = 10$

14. A certain amount of water is in a tank. What fraction of the tank is full?

    (1) If k more gallons of water are added, $\frac{5}{7}$ of the tank is full.
    (2) If k gallons are removed, $\frac{4}{7}$ of the tank is empty.

15. If $a = 3^{3x} + 1$, what is the value of a?

   (1) $3^{-x} = \frac{1}{2}$

   (2) $27^x = 8$

16. What is the ratio of this year's sales to last year's sales at store Q?

   (1) Store Q's sales were \$15 million last year.

   (2) The sales increased 30% compared with last year.

17. After receiving his salary, Tavares spent $\frac{1}{3}$ of it the first day and kept spending $\frac{1}{3}$ of the remaining amount each day. How much was his salary?

   (1) \$800 was left at the end of 4 days.

   (2) He spent \$900 on the second day.

18.

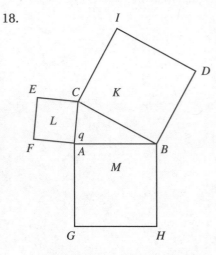

   $CIDB$, $ABHG$, and $ECAF$ are squares. Their areas are K, M, and L, respectively. Is q greater than 90°?

   (1) $K > L + M$

   (2) $CB - AB < AC$

19. Is x negative?

   (1) $x^2y < 0$

   (2) $x - y > 0$

20. Is $4a + b$ odd?

   (1) b is even.

   (2) a is odd.

21. What is the sum of $x + y + z = $?

   (1) $\frac{x}{3} = \frac{y}{5} = \frac{4}{z}$

   (2) $x \cdot z - 12 = 0$

22. What is the value of $\frac{x^2 - y^2}{xy} = $?

   (1) $\frac{x}{1} = \frac{y}{3}$

   (2) $x = 15$

23. 5, 2, and x are the lengths of the sides of triangle CBA. If x is an integer, what is its value?

    (1) x is prime.
    (2) CBA is an isosceles triangle.

24. m, n, and k are positive integers, and $\frac{m + 2n}{2} = 3k$. Is k an even integer?

    (1) m is even.
    (2) n is odd.

25.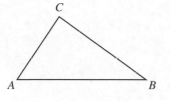

    Is AB longer than BC?

    (1) $m\angle A = 46°$ and $m\angle B = 43°$
    (2) $m\angle C = 91°$

26. If ab and ba are two-digit numbers where a and b represent digits, what is the value of $a - b$?

    (1) $ab - ba = a - b$
    (2) $a = 9$

27. In the equation $2x^2 + (k - 2)x - k + 4 = 0$, x is a variable and k is a constant. What is the value of k?

    (1) $x = \frac{1}{2}$ is a root of the equation.
    (2) $2x^2 + (k - 2)x - k + 4 = 0$ is divisible by $(x + 3)$.

28. x and y are two nonzero real numbers. Is $x > y$?

    (1) $y^2 - x^2 > 0$
    (2) $\frac{1}{x} - \frac{1}{y} < 0$

29. If $a^2 \le a$, what is the value of a?

    (1) $2a + 3$ is an integer.
    (2) $a^3 \le a^2$

30. Is the median age at the family picnic 50?

    (1) The oldest person at the picnic is 85, and the youngest person is 15.
    (2) Half of the group are older than Jerry, who is 50 years old.

31. In a particular high school, 60% of the teachers have graduate degrees. Is the number of female teachers with graduate degrees more than the number of male teachers with graduate degrees?

    (1) 40% of the teachers are male.
    (2) There are 18 female teachers.

32. If k is a positive integer, is $2^k + k^2$ odd?

    (1) $6k^2 + 2$ is even.
    (2) $5^k + k^4 + 3$ is odd.

33. m and n are two consecutive integers where $n > m$. What is the value of $m^2 - n^2$?

    (1) $m + n = 35$
    (2) $m - n = -1$

34. What is the value of $\frac{(a^2 c + cb)}{ca}$?

    (1) $\frac{a}{b} = 3$
    (2) $a = 5$

35. $(2^x - 3)^m = 1$. If m is a nonzero integer, what is the value of x?

    (1) m is odd.
    (2) x is even.

36. k, l, and m are three lines on the same plane. Is k parallel to l?

    (1) k is perpendicular to m
    (2) m is perpendicular to l

37. Amir, Bianca, and Carlos shared a certain number of questions. Bianca received how many more questions than Carlos?

    (1) Amir and Bianca received 24 questions in total.
    (2) Amir and Carlos received 20 questions in total.

38. There are only red and blue marbles in a hat. If a marble is to be picked at random, what is the probability that it will be red?

    (1) The number of red marbles is $\frac{4}{7}$ of the number of blue marbles.
    (2) The number of blue marbles is 30 more than the number of red marbles.

39. Is $a > 3$?

    (1) $-a + 2 < -1$
    (2) $7 - 2a < 1$

40. A law firm has two different hourly rates, one for junior lawyers and one for senior lawyers. Was the average hourly rate on case Q greater than $600?

    (1) The firm had 3 senior lawyers and 7 junior lawyers work on the case.
    (2) The senior lawyer rate is $750/hr, and the junior lawyer rate is $400/hr.

41. K and T are positive integers. When 248 is divided by K, the quotient is T and the remainder is 10. What is the value of K?

    (1) K is odd.
    (2) K is a three-digit integer.

42. Is $y > x$?

    (1) $(1 - x)(2 - y) < 0$
    (2) $x < 0$

43. What is the value of $\frac{x}{y}$?

    (1) $x - \frac{1}{y} = 2$

    (2) $y - \frac{1}{x} = 3$

44. *abcd* is a four-digit integer where *a*, *b*, *c*, and *d* represent different nonzero integers. Is $a + b + c + d > 11$?

    (1) $abcd > 6,000$

    (2) *abcd* is divisible by 5 and $8c = 4b = a$

45. ✪ is an operation defined in real numbers as $a✪b = a^2 + b^2 - 2ab$. What is the value of *a*?

    (1) $a✪b = 0$

    (2) $a = 4 - b$

46. What is the value of $x + y$?

    (1) $x^2 + y^2 = 17$ and $x \cdot y = 4$

    (2) $x - y = 3$ and $x^2 - y^2 = 15$

47. When *a* is divided by *b*, the quotient is 5 and the remainder is *k*. What is the remainder when 2*a* is divided by 5?

    (1) $k = 3$

    (2) $b = 8$

48. What is the value of *y*?

    (1) $y = 7 - 2x$

    (2) $x = \dfrac{14 - 2y}{4}$

49. What is the area of rectangle *ABCD*?

    (1) Perimeter = 24

    (2) The lengths of *AB* and *BC* are prime numbers.

50. Does $ab + bc + ac = \dfrac{5abc}{4}$?

    (1) $\dfrac{16}{a} + \dfrac{16}{b} + \dfrac{16}{c} = 20$

    (2) $4 + \dfrac{4c}{a} + \dfrac{4c}{b} = 5c$

51. *A* is a 15-digit integer. Is *A* divisible by 9?

    (1) Each digit of *A* is divisible by 3.

    (2) The sum of all digits of *A* is 72.

52. All cookies in a coffee shop contain nuts, chocolate chips, or both. If there are a total of 90 cookies, how many of them have nuts only?

    (1) The number of cookies with chocolate chips only is equal to the number of cookies with both.

    (2) The number of cookies with nuts only is three times the number of cookies with chocolate chips only.

53. What is the value of n?

(1) $2n + 2 \geq 10$

(2) $2 - \dfrac{n}{2} \geq 0$

54. What is the ratio of m to n if m and n are nonzero numbers?

(1) $16n^2 - 2m^2 = 7m^2$

(2) $4mn = 3m^2$

55. Two cars enter a 6-mile tunnel from each end at the same time. How many seconds after they enter do they meet?

(1) The sum of their speeds is 120 mi/hr.

(2) They meet at a point 4 miles from one end of the tunnel.

56.

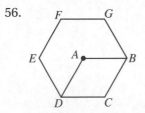

$BCDEFG$ is a regular hexagon, and A is the midpoint of EB (not shown). What is the area of $ABCD$?

(1) $AB = 6$

(2) $FD = 6\sqrt{3}$

57. Is a greater than 8?

(1) a minus b is not more than 7.

(2) The sum of a and b is less than or equal to 9.

58. Both Martha and Pete received a raise last year. Does Martha make more than Pete this year?

(1) Martha's raise was 6%, and Pete's raise was 8%.

(2) Both Martha and Pete received a raise of $4,800.

59.

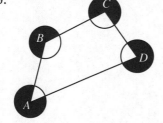

In the figure, all circles are identical and they are centered on the vertices of quadrilateral $ABCD$. What is the sum of the shaded areas?

(1) The radius of each circle is 8.

(2) $AB = BC = CD = \dfrac{2}{3} AD$

60. How many positive factors does m have?

    (1) m is the product of 3 prime numbers: a, b, and c.

    (2) m is even and divisible by 15.

61.

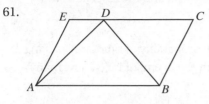

    If $ABCE$ is a parallelogram, what is the area of $\triangle ABD$?

    (1) D is the midpoint of EC.

    (2) The area of $ABCE$ is 60.

62. If x is a real number, is $x^2 + x$ positive?

    (1) $x > -1$

    (2) $x < 0$

63. x is a nonzero real number. Is x even?

    (1) $x^2 + x$ is even.

    (2) $\dfrac{x^2 + 6}{x}$ is an integer.

64. Does $\dfrac{2m}{n} = \dfrac{1}{5}$ if $n \neq 0$?

    (1) m is 10% of n.

    (2) $10^{-15} \cdot m = 10^{-16} \cdot n$

65.

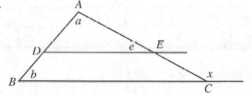

    What is the value of x if $DE \parallel BC$?

    (1) $a + b = 110°$

    (2) $e = 70°$

66. What is the value of $m^2 + 4mn + 4n^2 + \dfrac{m + 2n}{3} - 2m - 4n$?

    (1) $m + 2n = 6$

    (2) $\dfrac{m}{2n} = 2$

67. What is the slope of line ℓ_1?

    (1) Line $3ax + 4ay = 3$ is parallel to ℓ_1.

    (2) Line $2ax + 5y = 5a$ is perpendicular to ℓ_1.

68. Is $\dfrac{1}{a} > \dfrac{1}{b}$?

    (1) a and b are negative numbers.

    (2) $|b| > |a|$

69. Sergei owns 7 aquariums and a various number of goldfish in each of them. What is the average number of goldfish per aquarium?

   (1) The median number of goldfish in all aquariums is 6.
   (2) Four of the aquariums have the same number of goldfish, and the rest each have double the number of fish.

70. There are only green and yellow marbles in a bag. If the probability of randomly picking a yellow marble first and a green marble second without replacement is $\frac{4}{15}$, how many yellow marbles are there?

   (1) The ratio of yellow marbles to green marbles is 3 : 2.
   (2) There are two more yellow marbles than green marbles.

SOLUTIONS

1	2	3	4	5	6	7	8	9	10	11	12	13	14	15
C	E	D	A	A	C	C	E	E	C	C	E	B	C	D

16	17	18	19	20	21	22	23	24	25	26	27	28	29	30
B	D	A	E	A	E	A	D	E	D	A	D	E	E	B

31	32	33	34	35	36	37	38	39	40	41	42	43	44	45
E	B	A	C	D	C	C	A	D	E	C	C	C	D	C

46	47	48	49	50	51	52	53	54	55	56	57	58	59	60
B	A	E	C	D	B	C	C	B	A	D	C	C	A	A

61	62	63	64	65	66	67	68	69	70					
B	C	E	D	D	A	A	C	C	A					

1. **C** (1) a and b could be any two consecutive integers. NOT sufficient

   (2) $a^2 + b^2 = 613$ is not enough to find $a \cdot b$. For example, $(a + b)^2 = a^2 + 2ab + b^2$, but you need the value of $a + b$ to solve that equation. NOT sufficient.

   Using (1) and (2) together, you can write $a = b + 1$ from (1) and plug that into $a^2 + b^2 = 613$ from (2). $(b + 1)^2 + b^2 = 613$ can be solved for b and a can be found through (1).

   The answer is C; both statements together are sufficient.

2. **E** (1) If PR and RP are two-digit numbers, their values can be represented as follows;

   $PR = 10P + R$
   $RP = 10R + P$

   $10P + R + 10R + P = 11P + 11R = 132$
   $11(P + R) = 132$
   $P + R = 12$ does not allow us to solve for P. NOT sufficient

   (2) $P > R$ by itself does not provide enough information. NOT sufficient

When taken together, (1) states
$P + R = 12$ and (2) states $P > R$. Therefore PR could be 75 or 84.

(2) This states that $P > R$, but that does not restrict the possibilities to just one. PR can be still 75 or 83.

The answer is E; both statements together are still not sufficient.

3. **D** (1) From the division write $k = t \cdot 6 + 4$. So if t is given, k can be calculated easily. SUFFICIENT

(2) If $k = 6t + 4$ and $k = 2t + 20$ are given, the two equations can be solved simultaneously. SUFFICIENT
There is no need to solve for k, but it would look like.

$6t + 4 = 2t + 20$
$4t = 16$
$t = 4$

The answer is D; each statement alone is sufficient.

4. **A** First simplify the given equation by multiplying both sides by 4.
$4\left(k + \dfrac{t}{4}\right) = 3.75 \cdot 4$
$4k + t = 15$

Since k and t are positive integers, a limited number of (k, t) pairs work for the equation $4k + t = 15$. These are (1,11), (2,7), and (3,3).

(1) This states that $k < 5$ and $t < 5$.
$k = 3$, $t = 3$ is the only solution. SUFFICIENT

(2) Since the possible pairs of (k, t) are (1,11), (2,7), and (3,3), knowing that t is odd means t can be 11, 7, or 3. NOT sufficient

The answer is A; statement (1) alone is sufficient.

5. **A** (1) If $x^2 z < 0$, you can conclude that $z < 0$ since x^2 will always be positive and the product of x^2 and z must be negative. Also, from the question, you know that $x \cdot y < 0$. If the product of two nonzero numbers is negative, their quotient must be negative as well. Therefore $\dfrac{xz}{y} = \dfrac{x}{y} \cdot z$ must be $(-)(-) = (+)$

The answer to the question is no. SUFFICIENT

(2) If $yz < 0$, you cannot conclude if y and z are negative or positive. Without any information about z, you cannot decide if $\dfrac{xz}{y}$ is negative or positive. NOT sufficient

The answer is A; statement (1) alone is sufficient.

6. **C**

$$\overset{\hspace{4.5em}6}{\underset{\hspace{3.5em}P\hspace{3.5em}R}{\rule{11em}{0.4pt}}}$$

(1) Q can be on either side of R and still be 2 away from R. NOT sufficient

(2) The exact location of Q is not provided. NOT sufficient

Using (1) and (2) together, $PQ = 4$.

$$\overset{\hspace{2.5em}4\hspace{1.5em}2}{\underset{\hspace{1.5em}P\hspace{2em}Q\hspace{1.5em}R}{\rule{11em}{0.4pt}}}$$

The answer is C; both statements together are sufficient.

7. **C** (1) Ratios alone are not enough to conclude Clive's age. The ages could be $2:3:5$ or $4:6:10$ or any multiple of these. NOT sufficient

(2) Only the relationship between Deniz and Kaya is given. NOT sufficient

By using (1) and (2) together, you can set up equations to solve for all ages. Equations are given here for reference. You do not need to solve them for the purposes of this question.

$$\frac{D}{2} = \frac{K}{3} = \frac{C}{5}$$

Since $D = K - 5$ from (2), replace it in equation (1).

$$\frac{K-5}{2} = \frac{K}{3}$$
$$3K - 15 = 2K$$

Solve for K.

The answer is C; both statements together are sufficient.

8. **E** (1) No information is given about the number of correct, wrong, and blank answers. NOT sufficient

(2) You can set up $C = W + 5$. You cannot solve this equation since the total number of questions and the number of blank answers are unknown. NOT sufficient

By using (1) and (2) together, there's still no way of finding out how many of each the student answered since there is no indication about blank answers or about another equation between correct and wrong answers.

The answer is E; both statements together are still not sufficient.

9. **E** (1) The sum does not provide enough information to find any of the numbers. NOT sufficient

(2) An infinite number of different sets of even numbers can add up to 132. NOT sufficient

Using both (1) and (2) still does not provide the relative size of the 7 numbers.

(1) and (2) could have provided enough info if (2) had said that the 7 numbers are consecutive. However, that is not given.

The answer is E; both statements together are still not sufficient.

10. **C** (1) If $\ell_1 \parallel \ell_2$, $x + y = 180°$, which is not enough to solve for y. NOT sufficient

(2) $y = x + 40$. However, we are not given that $\ell_1 \parallel \ell_2$. NOT sufficient

By using (1) and (2) together, we can conclude $x + y = 180$ and $y = x + 40$, which can be solved together to find y.

The answer is C; both statements together are sufficient.

11. **C** (1) A percentage (ratio) relationship between A and B is given. Not enough information is given to relate C to A. NOT sufficient

(2) A percentage relationship between B and C is given. Not enough information is given to relate C to A. NOT sufficient

By using (1) and (2) together, the percentage relationship between A and C can be calculated.

(1) $A = 0.2B$
(2) $B = 0.3C$, plug this into (1)
   $A = 0.2(0.3)C$
   $A = 0.06C$
A is 6% of C

The answer is C; both statements together are sufficient.

12. **E** (1) The statement does not provide information about male to female ratio. NOT sufficient

(2) The statement does not provide information about the doctor to nurse ratio. NOT sufficient

By using (1) and (2) together, you can write:

$\dfrac{N}{D} = \dfrac{4}{1}$
$4D = N$
$\dfrac{M}{F} = \dfrac{1}{2}$
$M = 2F$

Even with the total of 150 doctors and nurses, it is not possible to solve for female doctors since the ratio of female doctors to male doctors can be different than the ratio of female nurses to male nurses. Only when put together is the ratio 1 : 2.

The answer is E; both statements together are still not sufficient.

13. **B** Multiply all terms of the first equation by z to simplify the denominator:

$z \cdot (xy) = z \cdot (6) - z \cdot \left(\frac{x}{z}\right)$
   $xyz = 6z - x$

(1) If $xyz = 20$, plug that in: $20 = 6z - x$

In this equation, it is not possible to isolate $\frac{z}{x}$. NOT sufficient

(2) If $y \cdot z = 10$ you can plug in 10 for $y \cdot z$.

$x \cdot 10 = 6z - x$
Add x to both sides.
$11x = 6z$
$\dfrac{z}{x} = \dfrac{11}{6}$

SUFFICIENT

The answer is B; statement (2) alone is sufficient

14. **C** (1) Let the initial amount of water in the tank be t and the total capacity of the tank be A. Statement (1) translates into:

$t + k = \dfrac{5}{7}A$

Although you only need the ratio $\frac{t}{A}$, you cannot get that from this equation alone. NOT sufficient

(2) This translates into $t - k = \frac{3}{7}A$, which cannot be solved for $\frac{t}{A}$ either. NOT sufficient

Using (1) and (2) together:

$$t + k = \frac{5}{7}A$$
$$t - k = \frac{3}{7}A \qquad \text{Add the equations to eliminate } k.$$
$$\overline{\phantom{2t = \frac{8}{7}A}}$$
$$2t = \frac{8}{7}A$$
$$\frac{t}{A} = \frac{4}{7}$$

The answer is C; both statements together are sufficient.

15. **D** 3^{3x} is equivalent to $(3^x)^3$. If you can get 3^x from the statements, then $3^{3x} + 1$ can be calculated.

(1) If $3^{-x} = \frac{1}{2}$, you can raise both sides to -1 power to get 3^x. $(3^{-x})^{-1} = \left(\frac{1}{2}\right)^{-1}$ and $3^x = 2$. SUFFICIENT

(2) If $27^x = 8$, then $(3^3)^x = 8$ and $3^{3x} = 8$. SUFFICIENT

The answer is D; each statement alone is sufficient.

16. **B** (1) This provides only a dollar amount but not a comparative measure between years. NOT sufficient

(2) Let the last year's sales be x. This year's sales become $x + 0.3x = 1.3x$. The ratio can be found by dividing this year's sales by last year's sales $\frac{1.3x}{x} = 1.3$. SUFFICIENT

The answer is B; statement (2) alone is sufficient.

17. **D** (1) If he spent $\frac{1}{3}$ of his money each day, $\frac{2}{3}$ of the beginning amount remains at the end of each day. Let his salary be S and set up (but do not solve) an equation in terms of the remaining amount:

$$S \cdot \frac{2}{3} \cdot \frac{2}{3} \cdot \frac{2}{3} \cdot \frac{2}{3} = 800$$

SUFFICIENT

(2) Let the salary be S again. So $\frac{2}{3}S$ remains after the first day. Tavares spent $\frac{2}{3} \cdot \frac{1}{3}S$ on the second day.

$$\frac{2}{3} \cdot \frac{1}{3} \cdot S = 900 \text{ can be solved for } S.\ \text{SUFFICIENT}$$

The answer is D; each statement alone is sufficient.

18. **A** (1) This states that $K > L + M$. Therefore $(CB)^2 > (AC)^2 + (AB)^2$. If ABC were a right triangle, $(CB)^2 = (AC)^2 + (AB)^2$ would be true. In this case, since $(CB)^2$ is greater, q must be greater than 90°. SUFFICIENT

(2) $CB - AB < AC$ means $CB < AC + AB$. In every triangle, any side is shorter than the sum of the other two sides. This statement does not provide any additional information. NOT sufficient

The answer is A; statement (1) alone is sufficient.

19. **E** (1) This does not provide any information about whether or not x is negative because x^2 is always positive. On the other hand, if $x^2y < 0$, y must be negative. NOT sufficient

(2) $x - y > 0$ means $x > y$, which does not provide any information about whether or not x is negative. NOT sufficient

By using (1) and (2) together, you know that x is greater than y and y is negative. However, that is not sufficient to find out if x is positive or negative. It could be either and still be greater than a negative y.

The answer is E; both statements together are still not sufficient.

20. **A** Regardless of a being odd or even, $4a$ will always be even. If you can tell whether b is even or odd, you can answer the question.

(1) b is even. SUFFICIENT

(2) As discussed above, knowing a is odd is not enough to answer the question. NOT sufficient

The answer is A; statement (1) alone is sufficient.

21. **E** (1) This provides information only about the ratios of x, y, and z. The statement does not help solve for x, y, z or for $x + y + z$. NOT sufficient

(2) This provides only the product of x and z. NOT sufficient

By using (1) and (2) together, there's still not enough information since the ratios of all variables are provided but no reference is given to their absolute sizes.

The answer is E; both statements together are still not sufficient.

22. **A** (1) Cross multiply to get $3x = y$. Replace $3x$ for y in the equation given in the question:
$$\frac{x^2 - y^2}{xy} = \frac{x^2 - (3x)^2}{x(3x)} = \frac{x^2 - 9x^2}{3x^2} = \frac{-8x^2}{3x^2} = \frac{-8}{3}$$
SUFFICIENT

(2) To find the required ratio, you need the value of y also or the ratio of y and x. NOT sufficient

The answer is A; statement (1) alone is sufficient.

23. **D** The sides of the triangle must satisfy the following inequality: $5 + 2 > x > 5 - 2$. So $7 > x > 3$. Therefore x could be $\{4, 5, 6\}$.

(1) Only 5 is prime among the eligible integers. SUFFICIENT

(2) If $x = \{4, 5, 6\}$ and CBA is an isosceles triangle, x must be 5. Note that x cannot be 2 since 2, 2, and 5 cannot be sides of a triangle. SUFFICIENT

The answer is D; each statement alone is sufficient.

24. **E** First rewrite the given equation as:
$$\frac{m + 2n}{2} - \frac{m}{2} + \frac{2n}{2} - \frac{m}{2} + n - 3k$$

(1) Knowing m is even is not enough information since $\frac{m}{2}$ could be either even or odd. For example, $\frac{4}{2}$ is even but $\frac{6}{2}$ is odd. NOT sufficient

(2) Knowing n is odd does not provide enough information since m is also needed to decide whether or not k is even. NOT sufficient

Even when (1) and (2) are taken together, a conclusion cannot be reached since $\frac{m}{2}$ remains unknown.

The answer is E; both statements together are still not sufficient.

25. **D** (1) AB is longer than BC if angle C is larger than angle A. Since the measures of angles A and B are given, the measure of angle C can be calculated. A, B, and C are interior angles of a triangle, and they add up to 180°. Since we can get the measure of angle C, we can answer the question. SUFFICIENT

    (2) If the measure of angle C is 91°, it must be the largest angle in the triangle since $A + B + C = 180°$. If you plug in 91° for C, $A + B + 91° = 180°$. So $A + B = 89°$. Neither A nor B can be greater than C. SUFFICIENT

    The answer is D; each statement alone is sufficient.

26. **A** If ab and ba are two-digit numbers, their values can be written as:

    $ab = 10a + b$
    $ba = 10b + a$

    (1) If $ab - ba = a - b$, replace the values from above into this equation:

    $$ab - ba = a - b$$
    $$10a + b - (10b + a) = a - b$$
    $$9a - 9b = a - b$$
    $$9(a - b) = (a - b)$$

    The equation can be true only if $(a - b) = 0$. SUFFICIENT

    (2) $a = 9$ by itself is not information to determine $a - b$. NOT sufficient

    The answer is A; statement (1) alone is sufficient.

27. **D** (1) If $x = \frac{1}{2}$ is a root, it can be plugged in for x and k can be determined. SUFFICIENT

    (2) If the equation is divisible by $(x + 3)$, that means $x = -3$ is a root of the equation. You can plug in -3 for x and solve for k. SUFFICIENT

    The answer is D; each statement alone is sufficient.

28. **E** (1) $y^2 - x^2 > 0$ means $y^2 > x^2$. This is not sufficient since x and y could both be negative, both be positive, or one of each. Note that $y = 3$ and $x = 2$, and $y = -3$ and $x = 2$ work with statement (1) but are conflicting results. NOT sufficient

    (2) $\frac{1}{x} - \frac{1}{y} < 0$ means $\frac{1}{x} < \frac{1}{y}$.

    Try $x = 3$ and $y = 2$:

    $x < y$, such as $\frac{1}{3} < \frac{1}{2}$

Try $x = -3$ and $y = 2$:

$x < y$, such as $-\frac{1}{3} < \frac{1}{2}$

NOT sufficient

Using (1) and (2) together is not sufficient to decide.

The answer is E; both statements together are still not sufficient.

29. **E** If $a^2 \leq a$, then a must be a number between 0 and 1. Only the squares of numbers between 0 and 1 are less than the original number. Notice that a^2 can equal a as well, so $0 \leq a \leq 1$

(1) If $2a + 3$ is an integer, a can be only 0, $\frac{1}{2}$, or 1. NOT sufficient

(2) $a^3 \leq a^2$ does not provide any additional information other than $0 \leq a \leq 1$. NOT sufficient

The answer is E; both statements together are still not sufficient.

30. **B** (1) Minimum and maximum values can be used to find the range but not the median. NOT sufficient

(2) The initial statement means that if you list all the ages from least to greatest, 50 will be in the middle. That is the definition of median. SUFFICIENT

The answer is B; statement (2) alone is sufficient.

31. **E** (1) This statement does not provide enough information on how graduate degrees are distributed among male and female teachers. You cannot assume that the 60% rate applies to both genders. NOT sufficient

(2) This statement does not provide additional information about the distribution of graduate degrees either. As discussed in (1), 60% is an overall rate. NOT sufficient

By using (1) and (2) together, you can conclude that 60% of the teachers are female and set up a proportion: $\frac{60}{100} = \frac{18}{x}$. Cross multiply and find $x = 30$, which is the number of teachers. The number of male teachers is 12. You can also find out that $60\% \cdot 30 = 18$ of the teachers have graduate degrees. However, there's still not enough information to calculate male and female graduate degree holders separately.

The answer is E; both statements together are still not sufficient.

32. **B** (1) $6k^2 + 2$ does not provide enough information about k since $6k^2 + 2$ is always even regardless of k. NOT sufficient

(2) If $5^k + k^4 + 3$ is odd, then $5^k + k^4$ must be even because even + odd = odd.

$5^k + k^4$ is even. All powers of odd numbers are odd, so 5^k must be odd. That makes k^4 odd since odd + odd is even.

If k^4 is odd, k must be odd since powers of only odd numbers are odd. SUFFICIENT

The answer is B; statement (2) alone is sufficient.

33. **A** (1) If m and n are consecutive integers and $m > n$, then $m - n = 1$. Since this statement provides $m + n = 35$, two equations can be solved simultaneously. SUFFICIENT.

(2) $m - n = -1$ does not provide new information since the question already indicates that m and n are consecutive numbers. NOT sufficient

The answer is A; statement (1) alone is sufficient.

34. **C** First simplify the expression in the question by splitting the denominator:

$$\frac{a^2 c + cb}{ca} = \frac{a^2 c}{ca} + \frac{cb}{ca} = a + \frac{b}{a}$$

(1) This provides $\frac{a}{b} = 3$ or $\frac{b}{a} = \frac{1}{3}$ but not the value of a. NOT sufficient

(2) The value b is also needed. NOT sufficient

By using (1) and (2) together, $\frac{b}{a} = \frac{1}{3}$ and $a = 5$ can be plugged in to get the value.

The answer is C; both statements together are sufficient.

35. **D** (1) If a power of $(2^x - 3)$ is 1 then $(2^x - 3)$ is either 1 or -1:

$(-1)^{\text{odd}} = -1$, $(-1)^{\text{even}} = 1$, $(1)^{\text{all powers}} = 1$:

If m is odd, then:

$(2^x - 3) = 1$
$\quad 2^x = 4$
$\quad\quad x = 2$

SUFFICIENT

(2) x has two solutions since $2^x - 3$ could be 1 or -1 depending on m.

$\begin{array}{ll} 2^x - 3 = 1 & \quad 2^x - 3 = -1 \\ \quad 2^x = 4 & \quad\quad 2^x = 2 \\ \quad 2^x = 2^2 & \quad\quad x = 1 \\ \quad\quad x = 2 & \end{array}$

If x is even, it must be 2. SUFFICIENT

The answer is D; each statement alone is sufficient.

36. **C** (1) Not enough information is given about l. NOT sufficient

(2) Not enough information is given about k. NOT sufficient

Using (1) and (2) together, if $k \perp m$ and $l \perp m$, k must be parallel to l since they are perpendicular to the same line on the same plane.

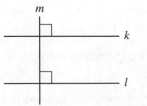

The answer is C; both statements together are sufficient.

37. **C** (1) Knowing how many Amir and Bianca received is not enough since we do not know the total number of questions or how many Carlos received. NOT sufficient

(2) Knowing how may Amir and Carlos received is not enough because of the same reasons. NOT sufficient

By using (1) and (2) together, you still cannot get the number of questions Amir, Bianca, and Carlos received individually. If you subtract the two equations side by side:

$$
\begin{array}{r}
A + B = 24 \\
- \quad A + C = 20 \\
\hline
B - C = 4
\end{array}
$$

Bianca received 4 more than Carlos.

The answer is C; both statements together are sufficient.

38. **A** The probability of picking a red marble when the hat contains only red (r) and blue (b) marbles is

$$\frac{r}{\text{total}} = \frac{r}{r + b}$$

(1) You can write $r = \frac{4}{7} \cdot b$ or $\frac{r}{4} = \frac{b}{7}$. This means $r = 4$ units and $b = 7$ units (or $r = 4x$ and $b = 7x$). The total is 11 units ($11x$).

$P(\text{red}) = \frac{4x}{11x} = \frac{4}{11}$. SUFFICIENT

(2) $b = 30 + r$ by itself is not sufficient to find $\frac{r}{r + b}$, since the total number of marbles is not provided. NOT sufficient

The answer is A; statement (1) alone is sufficient.

39. **D** (1) $-a + 2 < -1$ Add -2 to both sides.

         $-a < -3$ Multiply both sides by -1 and switch the inequality sign.

           $a > 3$ SUFFICIENT

   (2) $7 - 2a < 1$ Subtract 7 from both sides.

       $-2a < -6$ Divide by -2 and switch the inequality sign.

          $a > 3$ SUFFICIENT

The answer is D; each statement alone is sufficient.

40. **E** (1) The overall average rate cannot be calculated since the average rates of junior and senior lawyers are not given. NOT sufficient

(2) This does not provide the number of lawyers and hours spent. NOT sufficient

Using (1) and (2) together still does not provide enough information since we still do not know how many hours were spent by junior and senior lawyers. Note that you need the total hours for each since the hourly rate is required.

The answer is E; both statements together are still not sufficient.

41. **C** The remainder is 10. If you subtract the remainder from 248, the resulting number is divisible by K and T. $248 - 10 = 238$.

Alternatively, write $K \cdot T + 10 = 248$ and $K \cdot T = 238$. The factors of 238 are $2 \cdot 17 \cdot 7$.

1	238
2	119
7	34
14	17

(1) There is more than one odd integer. NOT sufficient

(2) Using the same analysis, notice that there are 2 three-digit integers (119 and 238) that could be K. NOT sufficient

By using (1) and (2) together, you can conclude that $K = 119$ since it is the only three-digit odd integer that divides 238.

The answer is C; both statements together are sufficient.

42. **C** (1) If $(1 - x)(2 - y) < 0$, then two cases are possible:

Case I: $(1 - x)$ is positive and $(2 - y)$ is negative.
$1 - x > 0$ and $2 - y < 0$
$1 < x$ and $2 > y$

Case II: $(1 - x)$ is negative and $(2 - y)$ is positive.
$1 - x < 0$ and $2 - y < 0$
$1 > x$ and $2 < y$
NOT sufficient

(2) $x < 0$ gives no information about y. NOT sufficient

When using (1) and (2) together if $x < 0$, then case I is not possible. Case II must be correct, which means $2 < y$, therefore $y > x$.

The answer is C; both statements together are sufficient.

43. **C** (1) Both sides of the equation can be multiplied by y to get $xy - 1 = 2y$. This equation alone is not enough to get $\frac{x}{y}$. NOT sufficient

(2) Both sides of the equation can be multiplied by x to get $xy - 1 = 3x$. This equation alone is not enough to get $\frac{x}{y}$. NOT sufficient

By using (1) and (2) together, $xy - 1 = 2y$ and $xy - 1 = 3x$. Since the left sides of the equations are equal, set the right sides equal to each other.

$2y = 3x$
$\frac{x}{y} = \frac{2}{3}$.

The answer is C; both statements together are sufficient.

44. **D** (1) If $abcd > 6{,}000$, a has to be 6, 7, 8, or 9. Although each digit cannot be identified, you know that a is at least 6. The other digits are at least 1, 2, and 3, so the sum is at least $6 + 1 + 2 + 3 = 12$. SUFFICIENT

(2) If $abcd$ is divisible by 5, d is either 5 or 0. It must be 5 since a, b, c, and d are nonzero.

If $8c = 4d = a$, c must be 1 since if you pick any number greater than 1, a and d become greater than 9.

If $a = 1$, $c = 8$, $d = 2$, then $a + b + c + d = 1 + 8 + 2 + 5 = 16$
SUFFICIENT

The answer is D; each statement alone is sufficient.

45. **C** If $a \otimes b = a^2 + b^2 - 2ab$, then
$a \otimes b = (a - b)^2$.

(1) If $a \odot b = 0$, then $(a - b)^2 = 0$.
So $a = b$, but this is not enough to find the value of a. NOT sufficient

(2) $a = 4 - b$ is not enough information to solve for a. NOT sufficient

By using (1) and (2) together, $a = b$ and $a = 4 - b$. Substitute the first equation into the second one:

$a = 4 - a$

$2a = 4$

$a = 2$

The answer is C; both statements together are sufficient.

46. **B** Remember that $(x + y)^2 = x^2 + 2xy + y^2$.

(1) Since you know $x^2 + y^2 = 17$ and $x \cdot y = 4$, plug them into the equation:

$(x + y)^2 = 17 + 2 \cdot 4 = 25$ Take the square root.

$x + y = \pm 5$ NOT sufficient

(2) Factor $x^2 - y^2$ as $(x - y)(x + y)$.
Since $x^2 - y^2 = (x - y)(x + y)$ and $x - y = 3$, plug them into the equation:
$15 = 3(x + y)$

$5 = x + y$ SUFFICIENT

The answer is B; statement (2) alone is sufficient.

47. **A** $a = 5b + k$

(1) If the remainder is 3:
$a = 5b + 3$
$2a = 2(5b + 3) = 10b + 6$

Since the remainder cannot be greater than 5 when dividing by 5, the remainder is $6 - 5 = 1$. SUFFICIENT

(2) If $b = 8$, then $a = 5 \cdot 8 + k = 40 + k$. Multiply each side by 2 to get $2a = 80 + 2k$. Without any information about k, the remainder cannot be determined. NOT sufficient

The answer is A; statement (1) alone is sufficient.

48. **E** (1) Not enough information is given to find the value of y. NOT sufficient

(2) Not enough information is given to find the value of y. NOT sufficient

Using (1) and (2) together:

$y = 7 - 2x$

$x = \dfrac{14 - 2y}{4}$ Multiply both sides by 4.

$4x = 14 - 2y$ Subtract 14 from both sides.

$4x - 14 = -2y$ Divide both sides by -2.

$y = 7 - 2x$ This is the same equation as (1).

Since both (1) and (2) provide the same information, you cannot find the value of y.

The answer is E; both statements together are still not sufficient.

49. **C** (1) Let the length be l and the width be w. $2(l + w) = 24$ means $l + w = 12$, which is not enough to find the area (area equals $l \cdot w$). NOT sufficient

(2) Not enough information is given to calculate the area. NOT sufficient

By using (1) and (2) together, $l + w = 12$ and l and w are prime numbers. The options for l and w are only 5 and 7. Area $= 5 \cdot 7 = 35$.

The answer is C; both statements together are sufficient.

50. **D** (1) Factor out 16 and make the denominators equal.

$$16\left(\frac{1}{a} + \frac{1}{b} + \frac{1}{c}\right) = 16\left(\frac{bc}{abc} + \frac{ac}{abc} + \frac{ab}{abc}\right) = 20$$

$$\frac{bc + ac + ab}{abc} = \frac{20}{16}$$

$$bc + ac + ab = \frac{5}{4}abc \qquad \text{Rearrange}$$

$$ab + bc + ac = \frac{5abc}{4} \qquad \text{SUFFICIENT}$$

(2) Multiply both sides by ab to eliminate the denominators.

$$4ab + \frac{4abc}{a} + \frac{4abc}{b} = 5abc$$

$$4ab + 4bc + 4ac = 5abc \qquad \text{Divide both sides by 4}$$

$$ab + bc + ac = \frac{5abc}{4} \qquad \text{SUFFICIENT}$$

The answer is D; each statement alone is sufficient.

51. **B** (1) To check if a number is divisible by 9, add all of its digits and see if the sum is divisible by 9. If the sum is divisible by 9, the number is also divisible by 9. Even though all digits are multiples of 3, their sum may not be a multiple of 9. For example, 14 digits with 3's and 1 digit with 6 add up to $14 \cdot 3 + 6 = 48$, which is not divisible by 9. If all the digits are 9's, the number would be divisible by 9. NOT sufficient

(2) If the sum of the digits is 72 which is $9 \cdot 8$, A must be divisible by 9. SUFFICIENT

The answer is B; statement (2) alone is sufficient.

52. **C** (1)

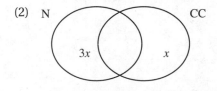

The first statement alone provides the information above. NOT sufficient

(2)

The second statement alone provides the information above. NOT sufficient

When (1) and (2) are taken together:

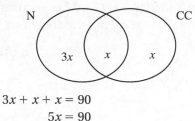

$$3x + x + x = 90$$
$$5x = 90$$
$$x = 18$$

Cookies with nuts only: $3x = 54$.

The answer is C; both statements together are sufficient.

53. **C** (1) $2n + 2 \geq 10$ Subtract 2 from each side.

           $2n \geq 8$ Divide each side by 2.

            $n \geq 4$ NOT Sufficient

     (2) $2 - \dfrac{n}{2} \geq 0$ Add $\dfrac{n}{2}$ to each side.

          $2 \geq \dfrac{n}{2}$ Multiply each side by 2.

          $4 \geq n$ NOT Sufficient

When using (1) and (2) together, $n \geq 4$ and $n \leq 4$. So n must be 4.

The answer is C; both statements together are sufficient.

54. **B** (1) $16n^2 - 2m^2 = 7m^2$ Add $2m^2$ to each side.

          $16n^2 = 9m^2$ Divide each side by $9n^2$.

          $\dfrac{16}{9} = \dfrac{m^2}{n^2}$ Take the square root of both sides.

$\dfrac{m}{n}$ could be $\dfrac{4}{3}$ or $\dfrac{4}{3}$. NOT sufficient

     (2) $4mn = 3m^2$ Divide both sides by $3mn$.

       $\dfrac{4mn}{3mn} = \dfrac{3m^2}{3mn}$

         $\dfrac{4}{3} = \dfrac{m}{n}$ SUFFICIENT

The answer is B; statement (2) alone is sufficient.

55. **A** (1) The time it takes two cars to meet is the same as the time it takes one car traveling at the sum of the speeds of the two cars to pass through the tunnel. If a car is traveling at a speed of 120 mi/hr:

$D = r \cdot t$

$6 = 120 \cdot t$

$t = \dfrac{6}{120}$ hrs   SUFFICIENT

(2) When they meet, the total distance traveled equals 6 miles:

$6 = r_1 \cdot t + r_2 \cdot t$

$r_1 \cdot t = 4$ and $r_2 \cdot t = 2$ are also given. We can divide these two equations side by side and get $\dfrac{r_1}{r_2} = 2$ but not the individual speeds. NOT sufficient

The answer is A; statement (1) alone is sufficient.

56. **D** (1) If you divide a regular hexagon into triangles as shown, each triangle becomes an equilateral triangle. Since $AB = 6$, each side equals 6. $ABCD$ covers 2 of these equilateral triangles, so its area is twice the area of one equilateral triangle. Find the area of an equilateral triangle using the formula $\frac{a^2\sqrt{3}}{4}$. SUFFICIENT

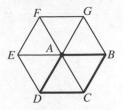

(2) FD is twice the height of one of the equilateral triangles that are shown in the figure. If the height of an equilateral triangle is known, its sides can be calculated using the ratios of the sides of a 30-60-90 triangle. Once the side length is known, the area can be calculated as well. SUFFICIENT

The answer is D; each statement alone is sufficient.

57. **C** (1) This statement means $a - b \leq 7$. NOT sufficient

(2) This statement means $a + b \leq 9$. NOT sufficient

By adding (1) and (2) together:

$$
\begin{array}{r}
a - b \leq 7 \\
+\ a + b \leq 9 \\
\hline
2a \leq 16 \\
a \leq 8 \qquad \text{SUFFICIENT}
\end{array}
$$

The answer is C; both statements together are sufficient.

58. **C** (1) Since the last year's salaries are not provided, this year's salaries cannot be calculated even though the raise percentage is given. NOT sufficient

(2) Knowing the dollar increase in salaries is not enough to calculate who is making more this year because last year's salaries are not known. NOT sufficient

By using (1) and (2) together, both Martha and Pete's salaries can be calculated.

For Martha, if 6% corresponds to a $4,800 increase, you can set up the proportion: $\frac{6}{100} = \frac{4,800}{x}$ to find her salary as of last year.

You can set up the proportion $\frac{8}{100} = \frac{4,800}{x}$ to find Pete's salary last year. Then compare.

The answer is C; both statements together are sufficient.

59. **A** (1) The shaded area in each circle has a central angle that is equal to 360° minus the corresponding interior angle of the quadrilateral. It is not possible to determine each angle individually. However, the sum of the interior angles of a quadrilateral is 360°. It is possible to find the sum of the central angles of the shaded regions as follows:

$4 \cdot 360° - 360° = 3 \cdot 360°$

Essentially, the shaded areas add up to 3 full circles. Since (1) provides the radius, the total area becomes $3 \cdot \pi \cdot 8^2$. SUFFICIENT

(2) The sides of the quadrilateral are irrelevant to the shaded areas. NOT sufficient

The answer is A; statement (1) alone is sufficient.

60. **A** (1) If m is the product of 3 prime numbers, $m = a \cdot b \cdot c$ and its factors are 1

1 and m
a and bc or   (1, m, a, b, c, ab, bc, ac)
b and ac
c and ba

Alternatively, pick 3 prime numbers and find their product. For example, use 2, 3, and 5. $m = 2 \cdot 3 \cdot 5 = 30$. The factors of 30 are 1, 2, 3, 5, 6, 10, 15, and 30. SUFFICIENT

(2) If m is even and a multiple of 15, it could be 30, 60, 90, and so on. NOT sufficient

The answer is A; statement (1) alone is sufficient.

61. **B** (1) This does not provide enough information since no dimensions or areas are given. NOT sufficient

(2) The area of $\triangle ABD$ is half of the area of $ABCE$ because they have the same base and the same height.

Area $\triangle ABD = \frac{1}{2} \cdot h \cdot AB$

Area $ABCE = h \cdot AB$

SUFFICIENT

The answer is B; statement (2) alone is sufficient.

62. **C** (1) Try several values for x.
Try 1: $1^2 + 1 = 2$ True
Try -0.5: $(-0.5)^2 + -0.5 = -0.25$ False

We get contradicting answers. NOT sufficient

(2) Try several values for x.

Try -2: $(-2)^2 + (-2) = 2$ True
Try -0.5: $(-0.5)^2 + -0.5 = -0.25$ False

We get contradicting answers. NOT sufficient

Using (1) and (2) together gives $-1 < x < 0$. So $x^2 + x$ will always be negative since x^2 will always be positive but its absolute value will be less than x, which is negative. For example:

$(-0.9)^2 + -0.9 = 0.81 - 0.9 < 0$

The answer is C; both statements together are sufficient.

63. **E** (1) Factor $x^2 + x = x(x + 1)$. This does not provide enough information.

If x is even, $x + 1$ is odd. So $x(x + 1)$ is even.

If x is odd, $x + 1$ is even. So $x(x + 1)$ is even. This means $x(x + 1)$ is always even regardless of x. NOT sufficient

(2) $\dfrac{x^2 + 6}{x} = \dfrac{x^2}{x} + \dfrac{6}{x} = x + \dfrac{6}{x}$

To make this expression an integer, x could be ± 1, ± 2, ± 3, or ± 6. NOT sufficient

Using (1) and (2) does not provide enough information either.

The answer is E; both statements together are still not sufficient.

64. **D** First simplify the question by cross multiplying, $10m = n$.

(1) Statement 1 states that $m = 0.1n$. If you multiply both sides by 10, you get $10m = n$. SUFFICIENT

(2) In $10^{-15} \cdot m = 10^{-16} \cdot n$, multiply both sides by 10^{16} and simplify.

$10^{16} \cdot 10^{-15} \cdot m = 10^{16} \cdot 10^{-16}n$

$\qquad\qquad 10m = n$

SUFFICIENT

The answer is D; each statement alone is sufficient.

65. **D** (1) The first statement allows you to calculate $\angle ACB$.

$a + b + \mathrm{m}\angle ACB = 180°$

$110° + \mathrm{m}\angle ACB = 180°$

$\qquad \mathrm{m}\angle ACB = 70°$

Since x and $\angle ACB$ add up to $180°$, x must be $110°$. SUFFICIENT

(2) Since $BC \parallel DE$, $\mathrm{m}\angle ACB$ must be the same as e, which is $70°$. Since x and $\angle ACB$ add up to $180°$, x must be $110°$. SUFFICIENT

The answer is D; each statement alone is sufficient.

66. **A** First simplify the question:

$m^2 + 4mn + 4n^2 = (m + 2n)^2$

$(m + 2n)^2 + \dfrac{m + 2n}{3} - 2(m + 2n)$

(1) Replace each $m + 2n$ with 6.

$6^2 + \dfrac{6}{3} - 2 \cdot 6$

SUFFICIENT

(2) $\dfrac{m}{2n} = 2$ means $m = 4n$. This does not give a numerical value for the expression given by itself. NOT sufficient

The answer is A; statement (1) alone is sufficient.

67. **A** In order to find the slope of ℓ_1, we need to find the slopes of the given lines first.

(1) Solve for y in $3ax + 4ay = 3$.

$$3ax + 4ay = 3 \qquad \text{Subtract } 3ax \text{ from both sides.}$$
$$4ay = -3ax + 3 \qquad \text{Divide both sides by } 4a.$$
$$y = \frac{-3}{4}x + \frac{3}{4a} \qquad \text{The slope is } \frac{-3}{4}.$$

The slope of ℓ_1 is also $-\frac{3}{4}$ since they are parallel. SUFFICIENT

(2) Solve for y in $2ax + 5y = 5a$.

$$2ax + 5y = 5a \qquad \text{Subtract } 2ax \text{ from both sides.}$$
$$5y = -2ax + 5a \qquad \text{Divide both sides by } 5.$$
$$y = \frac{-2a}{5}x + \frac{5a}{5} \qquad \text{The slope is } \frac{-2a}{5}.$$

Since we do not know the value of a, the slope of ℓ_1 cannot be determined. NOT sufficient

The answer is A; statement (1) alone is sufficient.

68. **C** (1) This statement does not indicate whether a or b is greater. NOT sufficient

(2) If $|b| > |a|$, $0 < a < b$ or $b < a < 0$. Try different numbers:

$$0 < 2 < 3 \qquad \frac{1}{2} > \frac{1}{3} \qquad \text{so} \quad \frac{1}{a} > \frac{1}{b}$$
$$-4 < -2 < 0 \qquad \frac{1}{-4} > \frac{1}{-2} \qquad \text{so} \quad \frac{1}{b} > \frac{1}{a}$$

The results are inconsistent. NOT sufficient

When using (1) and (2) together, there's only one possible case in which is $b < a < 0$. That is sufficient to answer the question.

The answer is C; both statements together are sufficient.

69. **C** (1) The median of a set does not indicate anything about the average of a set of numbers. NOT sufficient

(2) No actual numbers of fish are given. NOT sufficient

Use (1) and (2) together. If the median is 6 and 4 out of 7 aquariums have the same number of fish, those 4 aquariums must have 6 fish each to make the median 6. If 4 out of 7 numbers are the same, that number must be the median. Those 7 numbers could be arranged from least to greatest as follows:

$x\,x\,x\,x\,a\,b\,c$
$a\,x\,x\,x\,x\,b\,c$
$a\,b\,x\,x\,x\,x\,c$
$a\,b\,c\,x\,x\,x\,x$

In each case, the median is 6. So the rest of the aquariums have 12 fish each.

The average is $\dfrac{6 \cdot 4 + 12 \cdot 3}{7} = \dfrac{60}{7}$.

The answer is C; both statements together are sufficient.

70. **A** (1) The yellow marbles are Y, and the green marbles are G.

$$\frac{Y}{G} = \frac{3}{2}$$

Let Y be $3x$ and G be $2x$.

The probability of picking yellow first is $\frac{3x}{5x}$ and of picking green second without replacement is $\frac{2x}{5x-1}$. The probability of picking yellow first and green second is

$$\frac{3x}{5x} \cdot \frac{2x}{5x-1} = \frac{4}{15} \qquad \text{Multiply both sides by } \frac{5}{3}.$$

$$\frac{5}{3} \cdot \frac{3}{5} \cdot \frac{2x}{5x-1} = \frac{4}{15} \cdot \frac{5}{3}$$

$$\frac{2x}{5x-1} = \frac{4}{9} \qquad \text{Cross multiply.}$$

$$18x = 20x - 4$$

$$x = 2$$

The number of yellow marbles is $3x = 6$. SUFFICIENT

(2) Let G be t and Y becomes $t + 2$. The probability of picking yellow first is $\frac{t+2}{2t+2}$ and of picking green second without replacement is $\frac{t}{2t+1}$. The probability of picking yellow first and green second is

$$\frac{t+2}{2t+2} \cdot \frac{t}{2t+1} = \frac{4}{15}$$

$$\frac{t^2 + 2t}{4t^2 + 6t + 2} = \frac{4}{15} \qquad \text{Cross multiply.}$$

$$15t^2 + 30t = 16t^2 + 24t + 8$$

$$t^2 - 6t + 8 = 0 \qquad \text{Factor.}$$

$$(t-2)(t-4) = 0$$

$$t = 2 \text{ or } t = 4$$

There are two possible answers. NOT sufficient

The answer is A; statement (1) alone is sufficient.

Model Tests

7

ANSWER SHEET
Model Test 1

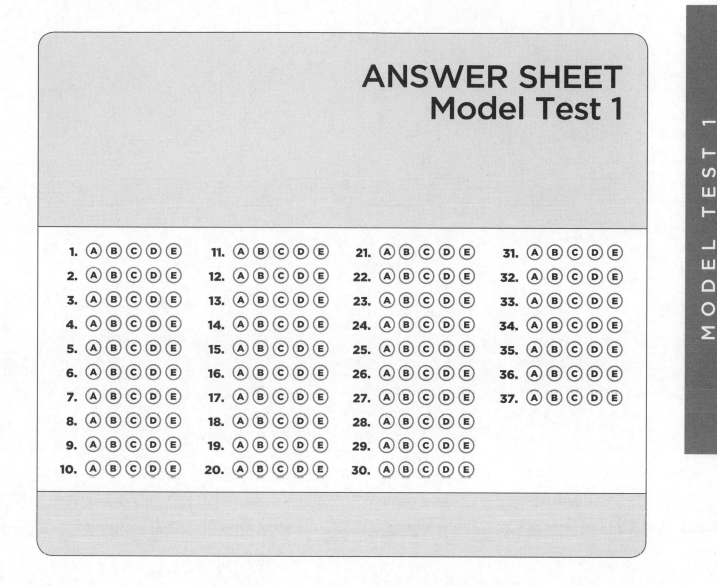

1. Ⓐ Ⓑ Ⓒ Ⓓ Ⓔ
2. Ⓐ Ⓑ Ⓒ Ⓓ Ⓔ
3. Ⓐ Ⓑ Ⓒ Ⓓ Ⓔ
4. Ⓐ Ⓑ Ⓒ Ⓓ Ⓔ
5. Ⓐ Ⓑ Ⓒ Ⓓ Ⓔ
6. Ⓐ Ⓑ Ⓒ Ⓓ Ⓔ
7. Ⓐ Ⓑ Ⓒ Ⓓ Ⓔ
8. Ⓐ Ⓑ Ⓒ Ⓓ Ⓔ
9. Ⓐ Ⓑ Ⓒ Ⓓ Ⓔ
10. Ⓐ Ⓑ Ⓒ Ⓓ Ⓔ

11. Ⓐ Ⓑ Ⓒ Ⓓ Ⓔ
12. Ⓐ Ⓑ Ⓒ Ⓓ Ⓔ
13. Ⓐ Ⓑ Ⓒ Ⓓ Ⓔ
14. Ⓐ Ⓑ Ⓒ Ⓓ Ⓔ
15. Ⓐ Ⓑ Ⓒ Ⓓ Ⓔ
16. Ⓐ Ⓑ Ⓒ Ⓓ Ⓔ
17. Ⓐ Ⓑ Ⓒ Ⓓ Ⓔ
18. Ⓐ Ⓑ Ⓒ Ⓓ Ⓔ
19. Ⓐ Ⓑ Ⓒ Ⓓ Ⓔ
20. Ⓐ Ⓑ Ⓒ Ⓓ Ⓔ

21. Ⓐ Ⓑ Ⓒ Ⓓ Ⓔ
22. Ⓐ Ⓑ Ⓒ Ⓓ Ⓔ
23. Ⓐ Ⓑ Ⓒ Ⓓ Ⓔ
24. Ⓐ Ⓑ Ⓒ Ⓓ Ⓔ
25. Ⓐ Ⓑ Ⓒ Ⓓ Ⓔ
26. Ⓐ Ⓑ Ⓒ Ⓓ Ⓔ
27. Ⓐ Ⓑ Ⓒ Ⓓ Ⓔ
28. Ⓐ Ⓑ Ⓒ Ⓓ Ⓔ
29. Ⓐ Ⓑ Ⓒ Ⓓ Ⓔ
30. Ⓐ Ⓑ Ⓒ Ⓓ Ⓔ

31. Ⓐ Ⓑ Ⓒ Ⓓ Ⓔ
32. Ⓐ Ⓑ Ⓒ Ⓓ Ⓔ
33. Ⓐ Ⓑ Ⓒ Ⓓ Ⓔ
34. Ⓐ Ⓑ Ⓒ Ⓓ Ⓔ
35. Ⓐ Ⓑ Ⓒ Ⓓ Ⓔ
36. Ⓐ Ⓑ Ⓒ Ⓓ Ⓔ
37. Ⓐ Ⓑ Ⓒ Ⓓ Ⓔ

1. $2x$ and $(y + 2)$ are inversely proportional. When $x = 1$, $y = 3$. What is the value of $8x^2$ when $y = 2$?

    (A) 10

    (B) 5

    (C) $\frac{5}{2}$

    (D) $\frac{25}{2}$

    (E) $\frac{25}{4}$

2. It takes 12 seconds to cut a uniform metal bar into four equal pieces. How long would it take to cut the same bar into 7 equal pieces?

    (A) 28 seconds

    (B) 24 seconds

    (C) 21 seconds

    (D) 18 seconds

    (E) 16 seconds

3. What is the value of $k + m$ if k and m are integers?

    (1) $k \cdot m^{-1} = 4$

    (2) $k^m = 64$

4. Spam filter A catches 90% of all spam e-mails sent to an e-mail box. Spam filter B catches only 75% of all spam e-mails. If Olivia uses both spam filters consecutively and the effectiveness of each filter stays constant under any scenario, what percent of the spam e-mails will make it into her e-mail account?

    (A) None

    (B) 1%

    (C) 2%

    (D) 2.5%

    (E) 5%

5. Barbara takes 3 hours to finish a task. The same task takes Ben 6 hours and Daria 9 hours to complete individually. Working at their constant rates, first, Ben and Barbara start working together and finish half of the task. Then Daria joins the two. They finish the rest of the task all together. How long did it take to finish the entire task?

    (A) $\frac{32}{11}$ hours

    (B) 2 hours

    (C) $\frac{21}{11}$ hours

    (D) $\frac{20}{11}$ hours

    (E) $\frac{19}{11}$ hours

6. Is $x = 3$?

    (1) $(x - 2)^{x+2} = 1$

    (2) $\left(\dfrac{x - 3}{x + 2}\right)^x = 0$

7. The sum of four consecutive odd integers is -96. Which of the following is the second largest number in the set?

    (A) -31

    (B) -29

    (C) -27

    (D) -25

    (E) -23

8. What is the ratio of r to s?

    (1) r is 32% more than s

    (2) r is 34 less than 2 times s

9. $\dfrac{\sqrt{6.25} + \sqrt{0.64} - \sqrt{0.09}}{\sqrt{92 - \sqrt{121}}} = ?$

    (A) $\dfrac{1}{9}$

    (B) $\dfrac{1}{3}$

    (C) 1

    (D) 3

    (E) 9

10. AB and BA are two-digit integers where A and B each represent a nonzero digit and $A > B$. Is AB divisible by 5?

    (1) $AB - BA$ is divisible by 9.

    (2) $AB + BA$ is divisible by 11.

11. It takes 10 workers 24 days to paint 8 identical buildings. Working at consistent rates, how many days would it take 6 workers to paint 4 of those buildings?

    (A) 18

    (B) 12

    (C) 18

    (D) 20

    (E) 24

12. MNK is the largest three-digit positive integer where $M \neq N \neq K$, and PQ is the largest two-digit negative integer where $P \neq Q$.

    $MNK - PQ = ?$

    (A) 910

    (B) 997

    (C) 999

    (D) 1,000

    (E) 1,098

13. When 64 is divided by $\frac{2x}{y}$, the quotient is $\frac{2x}{y}$ and the remainder is zero. What is the value of $\frac{x}{y} + 1$ if x and y are both positive integers?

(A) 9
(B) 8
(C) 7
(D) 5
(E) 4

14. A is the center of the circle shown. $ABCD$ is a square where CD and CB are tangent to the circle. What is the area of the circle?

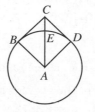

(1) The area of $ABCD$ is 12.
(2) The length of arc BED is $\pi\sqrt{3}$.

15. How many pounds of apples can Sydnie buy with all her money?

(1) She can buy either 8 pounds of apples and 6 pounds of pears or 2 pounds of apples and 10 pounds of pears with her money.
(2) Apples cost \$1.50 per pound, and pears cost \$1.30 per pound.

16. If $2^x + \frac{1}{2^{2-x}} = 80$, what is the value of x?

(A) 2
(B) 4
(C) 6
(D) 8
(E) 10

17. $3 \cdot \sqrt{3^3\sqrt{3^{-2}}} =$

(A) $\sqrt[3]{3}$
(B) $\sqrt{3}$
(C) 3
(D) $3\sqrt{3}$
(E) 9

18. The smallest of a set of consecutive integers is -12. If the sum of all integers in the set is 27, how many integers are in the set?

(A) 27
(B) 26
(C) 25
(D) 15
(E) 14

19. M dollars are split between two brothers, who are 15 and 16 years old, proportional to their ages. What is the value of M?

(1) The younger brother gets \$13 less.
(2) The older brother gets \$208.

20. For all real numbers, function ■ is defined as follows:

$$a \blacksquare b = \begin{cases} a^2 - b^2 & \text{if } a \geq b \\ a + b & \text{if } a < b \end{cases}$$

What is the value of $(4 \blacksquare 2) \blacksquare (2 \blacksquare 5)$?

(A) 5

(B) 14

(C) 15

(D) 19

(E) 95

21. A, B, C, D, E, F, and G are 7 consecutive odd integers. What is the average of these 7 integers?

(1) The average of D, E, F, and G is 64.

(2) $C + E = 122$

22. If $\frac{3}{4} < \frac{6}{n} + \frac{1}{2}$, what is the largest possible integer value of n?

(A) 22

(B) 23

(C) 24

(D) 25

(E) 26

23. Carl wants to buy 3 applications for his smart phone. If he picks 3 applications randomly among 6 games and 4 utility applications, what is the probability that he will get 2 games and 1 utility application?

(A) $\frac{1}{12}$

(B) $\frac{1}{6}$

(C) $\frac{3}{10}$

(D) $\frac{1}{2}$

(E) $\frac{2}{3}$

24. How many students are in the lab?

(1) If students are grouped so that there are 3 students per table, 4 students are left over.

(2) If students are grouped so that there are 4 students per table, there are 2 empty tables.

25. What is the value of $a \cdot e$?

(1) $a = 5b$; $\quad bd = 7$; $\qquad \frac{d}{e} = \frac{2}{3}$

(2) $a = \frac{5}{b}$; $\quad b = d + 7$; $\quad ed = 6$

26. 20% of a town's population is infected by a virus. If every week, 20% of the healthy people are getting infected, approximately what percent of the population will be infected by the end of the second week?

(A) 80%

(B) 60%

(C) 50%

(D) 40%

(E) 30%

27. A piggy bank contains Q quarters, D dimes, and N nickels. What is the ratio of the dollar amount of quarters to the total dollar value?

(A) $\dfrac{Q}{(D+N)}$

(B) $\dfrac{Q}{(Q+D+N)}$

(C) $\dfrac{25Q}{(Q+D+N)}$

(D) $\dfrac{25Q}{(10D+5N)}$

(E) $\dfrac{25Q}{(25Q+10D+5N)}$

28. How many days does it take to fill half of the pool with water?

(1) There are 200 gallons of water in the pool initially. Every day an amount equal to the existing water is added to the pool.

(2) The entire pool takes 10 days to fill.

29. A 39-year-old dad has two sons who are 3 years apart. Three years from now, the dad's age will be twice the sum of the ages of his sons. How old is the younger son today?

(A) 3 years old

(B) 6 years old

(C) 9 years old

(D) 12 years old

(E) 15 years old

30. If $t > 1$, is $m < 0$?

(1) $t - mt > 0$

(2) $m - tm > 0$

31. A document is k pages long with an average of k words per page. If Pat deletes a total of k words from the document without changing the number of pages, what is the new average number of words per page?

(A) k

(B) $k - 1$

(C) $k^2 - k$

(D) $\dfrac{k^2}{(k-1)}$

(E) $\dfrac{(k^2 - 1)}{k}$

32. p and r each represent a digit. Is $6prprp$ divisible by rp?

(1) $r = 6$

(2) $p = 0$

33. The ratio of the volumes of two spheres is s. What is the ratio of the radii of the two spheres? (Volume of a sphere: $V = \frac{4}{3}\pi r^3$)

(A) s^3

(B) s^2

(C) s

(D) $\sqrt{s}$

(E) $\sqrt[3]{s}$

34. If each side of a square is increased by 4 inches, the area of the square increases by 40. What is the area of the original square?

(A) 3

(B) 7

(C) 9

(D) 16

(E) 49

35. A 14-inch piece of wire is bent into a right triangle with a leg of 6 inches. What is the area of the triangle?

(A) $\frac{7}{4}$

(B) $\frac{21}{4}$

(C) $\frac{25}{4}$

(D) $\frac{75}{2}$

(E) $\frac{75}{4}$

36. If $y \neq 0$, is $x = 0$?

(1) $\dfrac{5x - y}{y} = 0$

(2) $3xy - 6x = 0$

37. Equilateral triangle ABC is placed on the rectangular coordinate system so that the coordinates of point A are $(-2, 0)$ and the coordinates of point B are $(4, 0)$. If the y-coordinate of point C is negative, how far is it from the origin?

(A) $-4\sqrt{3}$

(B) $2\sqrt{13}$

(C) $6\sqrt{3}$

(D) 12

(E) $8\sqrt{3}$

ANSWER KEY
Model Test 1

1. **D**	11. **D**	21. **D**	31. **B**
2. **B**	12. **B**	22. **B**	32. **A**
3. **C**	13. **D**	23. **D**	33. **E**
4. **D**	14. **D**	24. **C**	34. **C**
5. **D**	15. **A**	25. **A**	35. **B**
6. **B**	16. **C**	26. **C**	36. **A**
7. **E**	17. **E**	27. **E**	37. **B**
8. **A**	18. **A**	28. **C**	
9. **B**	19. **D**	29. **B**	
10. **E**	20. **E**	30. **B**	

ANSWERS TO MODEL TEST 1

1. **D** If x and y are inversely proportional, their product is constant:

   $$2x_1 \cdot (y_1 + 2) = 2x_2 \cdot (y_2 + 2)$$

   Divide by 2 and replace $x_1 = 1$, $y_1 = 3$, and $y_2 = 2$.

   (1) $(3 + 2) = x_2 (2 + 2)$

   $\qquad 5 = x_2 (4)$

   $\qquad \dfrac{5}{4} = x_2$

   $\qquad 8x_2^2 = 8 \cdot \left(\dfrac{5}{4}\right)^2 = 8 \cdot \dfrac{25}{16} = \dfrac{25}{2}$

2. **B** To divide a bar into 4 pieces, we need to make 3 cuts. If 3 cuts take 12 seconds, each cut takes 4 seconds.

   To cut the bar into 7 pieces, we need to make 6 cuts. That would take $6 \cdot 4 = 24$ seconds.

3. **C** (1) $k \cdot m^{-1} = k \cdot \dfrac{1}{m} = 4$

   Multiply each side by m, $k = 4m$. NOT sufficient

   (2) $k^m = 64$ is not enough information since $64^1 = 8^2 = 4^3 = 2^6 = 64$. NOT sufficient

   By using (1) and (2) together, we can substitute $k = 4m$ into k^m to get $(4m)^m = 64$. This equation will work only when $m = 2$.

   The answer is C; both statements together are sufficient.

4. **D** The first spam filter catches 90%, which means it lets in 10% of the spam. Similarly, filter B lets in 25% of the spam. In the end, 25% of 10% of all spam e-mails will make it into the inbox:

   $$25\% \cdot 10\% = 2.5\%$$

5. **D** If Barbara (x) and Ben (y) are working together:

   $$\dfrac{1}{x} + \dfrac{1}{y} = \dfrac{1}{t}$$

   $$\dfrac{1}{3} + \dfrac{1}{6} = \dfrac{2}{6} + \dfrac{1}{6} = \dfrac{3}{6} = \dfrac{1}{2}$$

   It takes them 2 hours to finish the entire task. So they spent 1 hour to finish one-half of the task. Once Daria (z) joins them and they start working together:

   $$\dfrac{1}{x} + \dfrac{1}{y} + \dfrac{1}{z} = \dfrac{1}{t}$$

   $$\dfrac{1}{3} + \dfrac{1}{6} + \dfrac{1}{9} = \dfrac{6}{18} + \dfrac{3}{18} + \dfrac{2}{18} = \dfrac{11}{18}$$

   It would take them $\dfrac{11}{18}$ hours to finish the entire task. To finish the second half of the task takes $\dfrac{1}{2} \cdot \dfrac{18}{11} = \dfrac{9}{11}$ hours.

   Add the two times:

   $$1 + \dfrac{9}{11} = \dfrac{11}{11} + \dfrac{9}{11} = \dfrac{20}{11} \text{ hours}$$

6.   **B** (1) If $(x-2)^{x+2} = 1$, then $x - 2 = 1$ or $x + 2 = 0$. Any number except 0 raised to the zero power equals 1.

$x - 2 = 1$

$\quad x = 3$

or

$x + 2 = 0$

$\quad x = -2$

or

$x - 2 = -1$

and $x + 2$ is even

NOT sufficient

(2) If $\left(\dfrac{x-3}{x+2}\right)^x = 0$, then $\dfrac{x-3}{x+2} = 0$. The only way $\dfrac{x-3}{x+2}$ could be zero is when $x - 3 = 0$ and $x = 3$. SUFFICIENT

The answer is B; statement (2) alone is sufficient.

7.   **E** The average of equally spaced numbers, such as consecutive odd integers, is the middle number.

$\dfrac{-96}{4} = -24$

Since we have consecutive odd integers, they must be:

$-27 \qquad -25 \qquad -23 \qquad -21$

Notice that -24 is in the middle of -25 and -23.

The second largest number is -23.

8.   **A** (1) r is 32% more than s means:

$r = 1.32 \cdot s \qquad$ Divide by s.

$\dfrac{r}{s} = 1.32 \qquad$ SUFFICIENT

(2) r is 34 less than 2 times s means:

$r = 2 \cdot s + 34 \qquad$ This cannot be solved for $\frac{r}{s}$. NOT sufficient

The answer is A; statement (1) alone is sufficient.

9.   **B** $\sqrt{6.25} = \sqrt{\dfrac{625}{100}} = \dfrac{25}{10} = 2.5$

$\sqrt{0.64} = 0.8$

$\sqrt{0.09} = 0.3$

$\sqrt{92 - \sqrt{121}} = \sqrt{92 - 11} = \sqrt{81} = 9$

$\dfrac{2.5 + 0.8 - 0.3}{9} = \dfrac{3}{9} = \dfrac{1}{3}$

10.   **E** (1) $AB - BA$ is always divisible by 9, regardless of the values of A and B.

$\begin{aligned} AB &= 10A + B \\ -\ BA &= 10B + A \\ \hline AB - BA &= 9A - 9B = 9(A - B) \end{aligned}$

AB could be 65 or 71. NOT sufficient

(2) $AB + BA$ is always divisible by 11, regardless of the values of A and B.

$$AB = 10A + B$$
$$+\ BA = 10B + A$$
$$\overline{AB + BA = 11A + 11B = 11(A + B)}$$

Again, AB could be 65 or 71. NOT sufficient

The answer is E; both statements together are still not sufficient.

11. **D** First calculate the total amount of days that 1 person would take to paint 8 buildings. It takes $24 \cdot 10 = 240$ days for 1 person to paint 8 buildings.

From here, you could either write a proportion or find the unit rate. If you calculate the rate, it takes $\frac{240}{8}$ days for 8 workers to paint 1 building. So 4 buildings would take $30 \cdot 4 = 120$ days for 1 person. It will take 6 workers $\frac{120}{6} = 20$ days to paint 4 buildings.

12. **B** The largest three digit positive integer where $M \neq N \neq K$ is 987.

The largest two-digit negative integer where $P \neq Q$ is -10.

$MNK - PQ = 987 - (-10) = 997$

13. **D** If the remainder is zero, 64 must be the product of the dividend and the quotient:

$$\frac{2x}{y} \cdot \frac{2x}{y} = \frac{4x^2}{y^2} = 64$$
$$\frac{x^2}{y^2} = 16$$

Take the square root of both sides.

$$\frac{x}{y} = \pm 4$$

Since x and y are both positive, $\frac{x}{y} = 4$.

$$\frac{x}{y} + 1 = 4 + 1 = 5$$

14. **D** (1) Notice that AB, which is one side of the square, is a radius of the circle. Since the first statement gives the area of the square, its side, which is the radius of the circle, can be calculated:

$$r = \sqrt{12} = 2\sqrt{3}$$

Then use the radius to find the area of the circle.

$$A = \pi r^2$$

SUFFICIENT

(2) Since $ABCD$ is a square, the center angle BAD is 90°. That means arc BED is $\frac{1}{4}$ of the circle. The circumference of the circle must be 4 times the length of the arc. $C = 4\pi\sqrt{3}$, which is equal to $2\pi r$. So since we know the radius, we can therefore calculate the area. SUFFICIENT

The answer is D; each statement alone is sufficient.

15. **A** Let Sydnie's total money be M, the price of apples per pound be A, and the price of pears per pound be P.

(1) $M = 8A + 6P$ and $M = 2A + 10P$. We can set these two equations equal to each other:

$$8A + 6P = 2A + 10P$$
$$6A = 4P$$
$$\frac{3A}{2} = P$$

Substitute this P into the first equation:

$$M = 8A + 6P = 8A + 6 \cdot \frac{3A}{2} = 17A$$

SUFFICIENT

(2) The prices of apples and pears alone are not enough information to answer the question. NOT sufficient.

The answer is A; statement (1) alone is sufficient.

16. **C** Multiply both sides by 2^{2-x} to simplify the fraction:

$$2^{2-x} \cdot 2^x + \frac{2^{2-x}}{2^{2-x}} = 2^{2-x} \cdot 80$$
$$2^2 + 1 = 2^{2-x} \cdot 80$$
$$\frac{5}{80} = 2^{2-x}$$
$$\frac{1}{16} = 2^{2-x}$$
$$2^{-4} = 2^{2-x}$$
$$-4 = 2 - x$$
$$x = 6$$

17. **E** Start by working the inner square root, $\sqrt{3^{-2}} = \sqrt{\frac{1}{9}} = \frac{1}{3}$. The overall equation becomes $3 \cdot \sqrt{3^3 \cdot \frac{1}{3}}$. Multiply the expressions under the square root to get $3\sqrt{3^2}$, which is equal to $3 \cdot 3 = 9$.

18. **A** Start by adding -12 to 12, -11 to 11, and so forth and realize that they each add up to zero:

$$\underbrace{-12 - 11 - 10 + \ldots + 11 + 12}_{\text{25 integers add up to zero}} \underbrace{+ 13 + 14}_{\text{2 integers}} = 27$$

The sum of all integers from -12 to 12 will be zero. Remember that zero is an integer between -12 and 12. There are $25 + 2 = 27$ integers.

19. **D** M dollars are split proportional to 15 and 16, which means the younger brother gets $15x$ and the older brother gets $16x$. $M = 15x + 16x = 31x$. If we can find x, we can answer the question.

(1) $16x - 15x = \$13$
$$x = \$13$$

SUFFICIENT

(2) $16x = \$208$

$\quad\quad x = \$13$

SUFFICIENT

The answer is D; each statement alone is sufficient.

20. **E** For (4 ■ 2), we use $a^2 - b^2$ since $4 \geq 2$.

$$(4 \text{ ■ } 2) = 4^2 - 2^2 = 12$$

For (2 ■ 5), we use $a + b$ since $2 < 5$.

$$(2 \text{ ■ } 5) = 2 + 5 = 7$$

The question then reduces to:

$$(12 \text{ ■ } 7) = 12^2 - 7^2 = 95$$

21. **D** (1) If the average of 4 of the consecutive odd integers is 64, their sum must be $64 \cdot 4 = 256$, which can be used to find each number. There is no need to set up an equation, but it would look like:

$$x + (x + 2) + (x + 4) + (x + 4) = 256$$

Alternatively, since the average is 64, the number in the middle (between E and F) must be 64. So the numbers are 61, 63, (64), 65, 67. SUFFICIENT

(2) In A, B, C, D, E, F, and G integers C and E are equidistant from D. This means the average of C and E is equal to D.

$$D = \frac{122}{2} = 61$$

D is also in the middle of the entire set. All the numbers are equally spaced because they are consecutive odd integers. So the average of the entire set is also $D = 61$. SUFFICIENT

The answer is D; each statement alone is sufficient.

22. **B** First, subtract $\frac{1}{2}$ from each side:

$$\frac{3}{4} - \frac{1}{2} < \frac{6}{n} + \frac{1}{2} - \frac{1}{2}$$

$$\frac{1}{4} < \frac{6}{n}$$

To compare the fractions easily, expand $\frac{1}{4}$ by 6 to make the numerators both 6. Notice that we cannot multiply both sides by n since we do not know if n is positive or negative.

$$\frac{6}{24} < \frac{6}{n}$$

The largest n could be is 23. If n were 24, the fractions would be equal and any integer greater than 24 would make the right side less than $\frac{1}{4}$.

23. **D** There are 3 different ways to get 2 games and 1 utility since the ordering does not matter:

GUG, GGU, or UGG. Let's calculate the probability of getting GUG:

$$\frac{6}{10} \cdot \frac{4}{9} \cdot \frac{5}{8} = \frac{1}{6}$$

Notice that probabilities for each scenario are the same. The denominators will always be the same (10, 9, 8), and the numerators will change only their order. So the overall probability is $3 \cdot \frac{1}{6} = \frac{1}{2}$.

Alternatively, if you prefer to use the combination formula:

$$\frac{(\text{Pick 2 out of 6}) \cdot (\text{Pick 1 out of 4})}{\text{Pick 3 out of 6: all possibilities}} = \frac{\binom{6}{2} \cdot \binom{4}{1}}{\binom{10}{3}}$$

$$\frac{\dfrac{6!}{(6-2)! \cdot 2!} \cdot \dfrac{4!}{(4-1)! \cdot 1!}}{\dfrac{10!}{(10-3)! \cdot 3!}} = \frac{\dfrac{6 \cdot 5 \cdot 4!}{4! \cdot 2} \cdot 4}{\dfrac{10 \cdot 9 \cdot 8!}{7! \cdot 3 \cdot 2}} = \frac{60}{120} = \frac{1}{2}$$

24. **C** (1) This does not give enough information since the number of tables is not given. One equation could be written; $S = 3t + 4$, where S is the number of students and t is the number of tables. NOT sufficient.

(2) This also does not give the number of tables. One equation could be written; $S = 4 \cdot (t - 2) = 4t - 8$. NOT sufficient.

By taking (1) and (2) together, set the equations equal to each other.

$S = 3t + 4$
$S = 4t - 8$
$3t + 4 = 4t - 8$
$t = 12$

If we know the number of tables, the number of students can be calculated.

The answer is C; both statements together are sufficient.

25. **A** (1) Start from the last equation and keep plugging in until all variables but a and e are eliminated:

$\dfrac{d}{e} = \dfrac{2}{3}$

$d = \dfrac{2e}{3}$ Plug into $bd = 7$

$b \cdot \dfrac{2e}{3} = 7$ Solve for b

$b = \dfrac{21}{2e}$ plug into $a = 5b$

$a = 5 \cdot \dfrac{21}{2e}$

$ae = \dfrac{105}{2}$

SUFFICIENT

(2) Similar to (1), start from the left:

$ed = 6$

$d = \dfrac{e}{6}$ Plug into $b = d + 7$

$b = \dfrac{6}{e} + 7$ Plug into $a = \dfrac{5}{b}$

$a = \dfrac{5}{\dfrac{6}{e} + 7} = \dfrac{5e}{6 + 7e}$

NOT sufficient

The answer is A; statement (1) alone is sufficient.

26. **C** Let the town population be 100;

At the beginning, $100 \cdot 0.2 = 20$ people are infected and 80 are not. During the first week, $80 \cdot 0.2 = 16$ more people get infected.

At the end of the first week, a total of $20 + 16 = 36$ people are infected. After one week, there are 64 healthy people left. During the second week, $64 \cdot 0.2 \doteq 12.8 \approx 13$ more people get infected.

At the end of two weeks, there are $36 + 13 = 49$ infected people, which is approximately 50%.

27. **E** Q quarters are worth $25Q$ cents.
D dimes are worth $10D$ cents.
N nickels are worth $5N$ cents.

The total value is $25Q + 10D + 5N$.

The ratio of the value of quarters to the total is

$$\frac{25Q}{(25Q + 10D + 5N)}$$

28. **C** (1) We do not know the capacity of the pool. NOT sufficient.

(2) This statement does not mention if the filling process adds the same amount every day or not. NOT sufficient.

By taking (1) and (2) together, the filling process is as follows:

Start   Day 1   Day 2   Day 3   Day 4
$200 +$ $200 +$ $400 +$ $800 +$ $1,600 + \ldots$

We can go up to Day 10 and find out the capacity of the pool. There is no need to get that number but we know we can.

The answer is C; both statements together are sufficient.

29. **B** Let the younger son be n years old today. The older son is $n + 3$ years old. Three years from now, they are all 3 years older.
$$39 + 3 = 2 \cdot (n + 3 + n + 3 + 3)$$
$$42 = 2(2n + 9)$$
$$42 = 4n + 18$$
$$24 = 4n$$
$$n = 6$$

The younger son is 6 years old today.

30. **B** (1) $t - mt > 0$ means $t \cdot (1 - m) > 0$.

The product of t and $(1 - m)$ is positive. t is already given to be positive, so $1 - m > 0$, which means $1 > m$. So m could be -1 or $\frac{1}{2}$. NOT sufficient.

(2) $m - tm > 0$ means $m \cdot (1 - t) > 0$.

The product of m and $(1 - t)$ is positive. t is already given to be greater than 1, so $(1 - t) > 0$ must be negative. Then m is negative since their product is positive. So $m < 0$. SUFFICIENT

The answer is B; statement (2) alone is sufficient.

31. **B** Initially the total number of words equals the number of pages times the number of words per page, $k \cdot k = k^2$. If Pat deletes k words, the new total will be $k^2 - k$.

Since the number of pages is still k, the new average will be $\dfrac{(k^2 - k)}{k} = \dfrac{k(k-1)}{k} = k - 1$.

32. **A** (1) If $r = 6$, the question becomes is $6p6p6p$ divisible by $6p$?

$\dfrac{6p6p6p}{6p} = 10101$

For example, $\dfrac{656565}{65} = 10101$.

Any digit you replace for p will give a quotient of 10101. SUFFICIENT

(2) If $p = 0$, the question becomes is $6 \cdot 0r0r0$ divisible by $r0$? Not necessarily, because for $r = 1$ the answer is yes and for $r = 7$, the answer is no. NOT Sufficient

The answer is A; statement (1) alone is sufficient.

33. **E** $V_1 = \dfrac{4}{3}\pi r_1^{\,3}$

$V_2 = \dfrac{4}{3}\pi r_2^{\,3}$

$\dfrac{V_1}{V_2} = \dfrac{\frac{4}{3}\pi r_1^{\,3}}{\frac{4}{3}\pi r_2^{\,3}} = s$ Simplify $\dfrac{4}{3}$ and π.

$\dfrac{r_1^{\,3}}{r_2^{\,3}} = s = \left(\dfrac{r_1}{r_2}\right)^3$ Take the cube root of both sides.

$\dfrac{r_1}{r_2} = \sqrt[3]{s}$

34. **C** Let one side of the original square be n.

Original area = n^2
Increased area = $(n + 4)^2$

$(n + 4)^2 = n^2 + 40$ Square the left side.
$n^2 + 8n + 16 = n^2 + 40$ Subtract n^2 and 16 from both sides.
$8n = 24$
$n = 3$

The original area is $n^2 = 3^2 = 9$.

35. **B** If one of the legs of a right triangle is 6, the sum of the other leg and the hypotenuse must be $14 - 6 = 8$. Let the second leg be n, and then the hypotenuse becomes $8 - n$. Use the Pythagorean theorem:

$6^2 + n^2 = (8 - n)^2$

$36 + n^2 = 64 - 16n + n^2$ Subtract n^2 from both sides.

$36 = 64 - 16n$

$16n = 64 - 36 = 28$

$n = \dfrac{28}{16} = \dfrac{7}{4}$

The second leg is $\dfrac{7}{4}$ inches long.

The area is $\dfrac{1}{2} \cdot \text{leg}_1 \cdot \text{leg}_2 = \dfrac{1}{2} \cdot 6 \cdot \dfrac{7}{4} = \dfrac{21}{4}$.

36. **A** (1) Split the fraction to get $\frac{5x}{y} - \frac{y}{y} = \frac{5x}{y} - 1 = 0$.

This means $\frac{5x}{y} = 1$. Therefore $x \neq 0$. The answer is no. SUFFICIENT

(2) $3xy - 6x = 3x(y - 2) = 0$. x could be zero or y could equal 2. This statement is inconclusive. NOT sufficient

The answer is A; statement (1) alone is sufficient.

37. **B** First sketch the points on the xy-coordinate system.

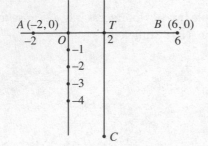

AB equals 8, which is one side of the equilateral triangle. Since the x-coordinate of C has to be in the middle of A and B, it must be 2. Draw TC in the figure to show the height of the equilateral triangle. Using the 30-60-90 triangle ratios or information from the geometry section, we can conclude that $TC = 4\sqrt{3}$. That means the coordinates of point C are $(2, 4\sqrt{3})$.

By using the Pythagorean theorem in triangle OTC or the distance formula, we can calculate the length of OC:

$$\sqrt{(4\sqrt{3})^2 + 2^2} = \sqrt{52} = 2\sqrt{13}$$

ANSWER SHEET
Model Test 2

1. (A) (B) (C) (D) (E)   11. (A) (B) (C) (D) (E)   21. (A) (B) (C) (D) (E)   31. (A) (B) (C) (D) (E)
2. (A) (B) (C) (D) (E)   12. (A) (B) (C) (D) (E)   22. (A) (B) (C) (D) (E)   32. (A) (B) (C) (D) (E)
3. (A) (B) (C) (D) (E)   13. (A) (B) (C) (D) (E)   23. (A) (B) (C) (D) (E)   33. (A) (B) (C) (D) (E)
4. (A) (B) (C) (D) (E)   14. (A) (B) (C) (D) (E)   24. (A) (B) (C) (D) (E)   34. (A) (B) (C) (D) (E)
5. (A) (B) (C) (D) (E)   15. (A) (B) (C) (D) (E)   25. (A) (B) (C) (D) (E)   35. (A) (B) (C) (D) (E)
6. (A) (B) (C) (D) (E)   16. (A) (B) (C) (D) (E)   26. (A) (B) (C) (D) (E)   36. (A) (B) (C) (D) (E)
7. (A) (B) (C) (D) (E)   17. (A) (B) (C) (D) (E)   27. (A) (B) (C) (D) (E)   37. (A) (B) (C) (D) (E)
8. (A) (B) (C) (D) (E)   18. (A) (B) (C) (D) (E)   28. (A) (B) (C) (D) (E)
9. (A) (B) (C) (D) (E)   19. (A) (B) (C) (D) (E)   29. (A) (B) (C) (D) (E)
10. (A) (B) (C) (D) (E)   20. (A) (B) (C) (D) (E)   30. (A) (B) (C) (D) (E)

1. 10% of an island's population is infected with a rare type of flu. A researcher discovered that each sick person infects two healthy people in one week. What percent of the island's population will be infected two weeks from now?

   (A) 40%
   (B) 60%
   (C) 90%
   (D) 99%
   (E) The entire island

2. What is the perimeter of the isosceles triangle ABC?

   (1) Two of its sides are 4 and 6 inches long.
   (2) Two of it angles are 50° and 80°.

3. In a rugby tournament with 12 teams, each team must play every other team once in the first round. How many total games are in the first round?

   (A) 11
   (B) 55
   (C) 66
   (D) 110
   (E) 111

4. When €0.642 was worth $1, what was the approximate value of $280 in euros?

   (A) €160
   (B) €180
   (C) €200
   (D) €400
   (E) €440

5. Andy is a years old, and his son Drew is d years old. In how many years (in terms of a and d) will Andy be 4 times as old as Drew?

   (A) $\dfrac{a - 4d}{4}$
   (B) $\dfrac{3a + d}{4}$
   (C) $\dfrac{a - 4d}{3}$
   (D) $\dfrac{a - d}{4}$
   (E) $\dfrac{4a - d}{3}$

6. Is integer P divisible by 84?

   (1) P is divisible by 6.
   (2) P is divisible by 14.

7. Eve and Greg together have $S. If Greg has $m more than Eve, how much money does Eve have in terms of S and m?

(A) $S - 2m$

(B) $\dfrac{(S - 2m)}{2}$

(C) $S - m$

(D) $\dfrac{(S - m)}{2}$

(E) $S + m$

8.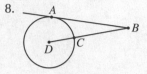

AB is tangent to the circle at A. What is the radius of the circle with center D?

(1) $AB = 24$

(2) $BC = 18$

9. If k cookies sell for b dollars, how many cookies can be purchased for m dollars?

(A) $\dfrac{mb}{k}$

(B) $\dfrac{bk}{m}$

(C) mbk

(D) $\dfrac{mk}{b}$

(E) $\dfrac{m}{bk}$

10. $\left(\dfrac{\sqrt{7}}{\sqrt{14}} + \dfrac{3\sqrt{3}}{\sqrt{6}} \right) \div \dfrac{\sqrt{16}}{\sqrt{2}} = ?$

(A) $\dfrac{\sqrt{2}}{4}$

(B) $\dfrac{\sqrt{2}}{2}$

(C) 1

(D) $\dfrac{\sqrt{6}}{2}$

(E) $\dfrac{2}{\sqrt{3}}$

11. What is the percent profit from the sale of one pair of shoes?

(1) The store buys 3 pairs for $5T and sells 2 pairs for $6T.

(2) The sales price of one pair of shoes is $\dfrac{9}{5}$ of the cost of one pair.

12. If $m < n < 0 < t$, which of the following must be false?

(A) $m^2 - n^2 > 0$

(B) $\dfrac{n}{t} + \dfrac{t}{n} < 0$

(C) $\dfrac{m}{n} + \dfrac{n}{m} > 0$

(D) $mn - mt < 0$

(E) $nt - nm < 0$

13. What is the value of $\frac{1}{a} + \frac{1}{b} + \frac{1}{c}$?

    (1) $a = 2b$ and $b = 3c$

    (2) $ab + ac + bc = 2abc$

14. If the terms $2t - 6$ and $t + 8$ represent two consecutive even integers, what is the greatest possible value of their sum?

    (A) 38

    (B) 42

    (C) 46

    (D) 50

    (E) 56

15. What is the value of $\frac{a^{-2} - b^{-2}}{a^{-1} - b^{-1}}$?

    (1) $\frac{a - b}{ab} = -\frac{2}{5}$

    (2) $a + b = \frac{3}{5}ab$

16. x and y are integers, and $-6 < x \leq 8$ and $-3 \leq y < 5$. The greatest possible value of $3x - 2y$ is how much greater than the least possible value of $3x - 2y$?

    (A) 63

    (B) 53

    (C) 48

    (D) 46

    (E) 37

17. The roots of the quadratic equation $x^2 - 4x + m - 3 = 0$ are x_1 and x_2. If $x_2 = 3x_1$, what is the value of m?

    (A) 4

    (B) 5

    (C) 6

    (D) 8

    (E) 9

18. b is what fraction of the sum of a and b?

    (1) The ratio of the sum of a and b to a is 6.

    (2) a is 20% of b.

19. If $5^{n+3} - 5^{n+1} = 3{,}000$, what is the value of n?

    (A) 1

    (B) 2

    (C) 3

    (D) 4

    (E) 5

20. One-third of the sum of two positive integers equals their difference. If the sum of the squares of the two numbers is 245, what is the difference of their squares?

(A) 7
(B) 14
(C) 98
(D) 147
(E) 174

21. Is $b - a$ negative?

(1) $a \cdot b < 0$
(2) $a - b > 0$

22. The cost of a purse to a retailer is $120. At what price should the retailer list the purse to make a 20% profit after offering a 20% discount?

(A) $144
(B) $168
(C) $172
(D) $180
(E) $196

23. On June 11th, Investor A had 2,400 shares of stock X and began selling those shares at a constant rate of 60 shares per day. On the same day, Investor B began buying stock X at a constant rate of 100 shares per day. If Investor B had 320 shares of stock X prior to his purchase on June 11th, after how many days did the two investors have the same number of shares of stock X?

(A) 10 days
(B) 11 days
(C) 12 days
(D) 13 days
(E) 14 days

24. There are k blue balls and t red balls in a bag. Two balls are picked consecutively without replacement. What is the probability of getting a blue ball in the first pick and a red ball in the second pick?

(A) $\dfrac{kt}{k+t}$

(B) $\dfrac{k}{k+t} + \dfrac{t}{k+t-1}$

(C) $\dfrac{k}{k+t} \cdot \dfrac{t}{k+t-1}$

(D) $\dfrac{k+t}{k+t-1}$

(E) $\dfrac{k}{k+t} \cdot \dfrac{t}{k+t}$

25. Is $\dfrac{b-1}{b-a}$ positive?

(1) $0 < a < 1$
(2) $ab > b$

26. What is the ratio of r to s?

   (1) $\frac{4}{r} + s = 18$

   (2) $\frac{4}{s} + r = 12$

27. There are twice as many girls as boys on a water polo team. If two players are picked at random, the probability of picking two girls is $\frac{22}{51}$. What is the probability of picking two boys at random?

   (A) $\frac{5}{51}$

   (B) $\frac{11}{51}$

   (C) $\frac{16}{51}$

   (D) $\frac{19}{51}$

   (E) $\frac{29}{51}$

28. At the end of 30 days, Ian made $2,600. How many days did he actually work?

   (1) For every day he worked, Ian made $140 net. For every day he did not, he spent $60.

   (2) The number of days he worked is 6 more than twice the number of days he did not work.

29. A senior consultant charges M dollars per hour for his services. His associate charges $\frac{2}{5}$ of that rate. If for a certain project the associate worked twice as many hours as the senior consultant, in terms of M what is the average hourly rate for the project?

   (A) $\frac{2M}{5}$

   (B) $\frac{3M}{5}$

   (C) $\frac{4M}{5}$

   (D) $\frac{9M}{10}$

   (E) $\frac{6M}{5}$

30. The cost of a cab ride in a certain city is t for the first mile and n¢ for each 0.2 miles after that. If Sandy paid P, how far was her ride in terms of n, t, and P?

   (A) $\frac{20}{n}(P - t)$

   (B) $\frac{P - t}{5n}$

   (C) $\frac{P - t}{5n} + 1$

   (D) $\frac{P - t}{5n} - 1$

   (E) $\frac{20}{n}(P - t) + 1$

31. Every day, Joe's daily wage increases by an amount equivalent to his total wages from the day before. How much does Joe earn on the 11th day?

    (1) He made $573.44 on the 14th day.

    (2) He made a total of $4.41 in the first 6 days.

32. Ronaldinho invests part of his $10,000 in an account earning 6% simple interest per year. He invests the rest in another account earning 7.5% simple interest per year. If after 3 years the total interest earned is $1,980, what is the ratio of the amount invested in 6% to the amount invested in 7.5%?

    (A) $\frac{3}{2}$

    (B) $\frac{4}{3}$

    (C) 1

    (D) $\frac{3}{4}$

    (E) $\frac{2}{3}$

33. A retail clothing store is planning on sending out questionnaires via e-mail. The marketing department estimates that 40% of all e-mails get discarded as spam and only 20% of the people who receive the e-mail actually start the survey. Finally, the department estimates only $\frac{1}{3}$ of the people who start the survey actually finish it. If the marketing department wants at least 360 completed responses, how many e-mails should be sent at minimum?

    (A) 3,000

    (B) 6,000

    (C) 9,000

    (D) 15,000

    (E) 18,000

34. In the figure, the circle is tangent to the y-axis with a center at O. The line that passes through D, O, and B (not shown) is parallel to the y-axis. If the coordinates of point B are (n, m), which of the following could be the coordinates of point D?

    (A) $(n, -m)$

    (B) $(n, n - m)$

    (C) $(n, 2n + m)$

    (D) $(n, n + m)$

    (E) $(n, 2n - m)$

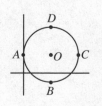

35. Is $\dfrac{k}{t} = \dfrac{m}{n}$?

    (1) $\dfrac{k}{t-k} = \dfrac{m}{n-m}$

    (2) $\dfrac{k}{t+k} = \dfrac{m}{n+m}$

36. The dimensions of rectangle $ABCD$ are x and y, and the dimensions of rectangle $EFGH$ are $(x + y)$ and $(x - y)$. What is the ratio of the diagonal of $EFGH$ to the diagonal of $ABCD$?

    (A) $\dfrac{x}{y}$

    (B) $\dfrac{x}{x + y}$

    (C) $\sqrt{2}$

    (D) $\dfrac{x}{x - y}$

    (E) $\dfrac{y}{x}$

37. If m and n each represent a nonzero digit, what is the value of $\dfrac{mn}{m.n} + \dfrac{mn}{0.mn} - \dfrac{mm}{10}$?

    (1) $n = 2$

    (2) $m = 3$

ANSWER KEY
Model Test 2

1. **C**	11. **D**	21. **B**	31. **D**
2. **C**	12. **D**	22. **D**	32. **A**
3. **C**	13. **B**	23. **D**	33. **C**
4. **B**	14. **D**	24. **C**	34. **C**
5. **C**	15. **B**	25. **C**	35. **D**
6. **E**	16. **B**	26. **C**	36. **C**
7. **D**	17. **C**	27. **A**	37. **B**
8. **C**	18. **D**	28. **D**	
9. **D**	19. **B**	29. **B**	
10. **C**	20. **D**	30. **E**	

1. **C** Assume there are 100 people on this island. That means 10 of them are infected in the beginning.

   10 will infect 20 in one week. At the end of 1 week, there will be 30 sick people.

   30 sick people will infect 60 people during the second week. At the end of the second week, there will be 90 sick people. Therefore, the percentage of sick people will be $\frac{90}{100} = 90\%$.

2. **C** (1) The triangle could have legs of either 4-4-6 or 4-6-6. NOT sufficient

   (2) The statement does not provide any side lengths. NOT sufficient

   Taking (1) and (2) together shows that two angles are 50° and 80°. This means the third angle is 50°. Then we can conclude that the side opposite the 50° angle is 4 since it is the shorter side. Therefore, the triangle must have legs of 4-4-6. The perimeter equals 14.

   The answer is C; both statements together are sufficient.

3. **C** Team 1 will play 11 games; Team 2 will play an additional 10 games since we've already counted its match with Team 1, and so on.

   Add $11 + 10 + 9 + 8 + 7 + 6 + 5 + 4 + 3 + 2 + 1$.

   For easy addition, rearrange as $1 + 9$, $2 + 8$, $3 + 7$, and $4 + 6$ to make four 10s. $11 + 10 + 10 + 10 + 10 + 10 + 5 = 66$ games

4. **B** Set up a proportion:

   $$\frac{€}{\$} \rightarrow \frac{0.642}{1} = \frac{x}{280}$$
   $$x = 280 \cdot 0.642$$
   $$x = €179.76$$
   $$x \approx €180$$

   Alternatively, use the process of elimination. 0.642 is slightly larger than 0.6.

   $0.6 \cdot \$280 = \$168$

   Your answer should be slightly more than $168. The closest answer is $180.

5. **C** Let the number of years be x. In x years, Andy will be $a + x$ years old and Drew will be $d + x$ years old. Set up the equation as follows:

   $$a + x = 4 \cdot (d + x)$$
   $$a + x = 4d + 4x$$
   $$a - 4d = 3x$$
   $$\frac{a - 4d}{3} = x$$

6. **E** (1) P could be 36, 168, or any number divisible by 6. However, 36 is not divisible by 84. NOT sufficient

   (2) P could be 28, 84, or any number divisible by 14. However, 28 is not divisible by 84. NOT sufficient

Taking (1) and (2) together is not sufficient because P could be 42, which is divisible both by 6 and 14 but not by 84. P could also be 84, which is divisible by 6, 14, and 84.

The answer is E; both statements together are still not sufficient.

7. **D** Let Eve's money be E. Then Greg's money becomes $E + m$. Set up the following equation:

$$E + E + m = S$$
$$2E + m = S$$
$$2E = S - m$$
$$E = \frac{(S - m)}{2}$$

8. **C** (1) AB alone does not provide enough information since we cannot create an equation for the radius. We are given AB, but we do not know the length of either BD or BC. We know that if we draw DA, ABD would be a right triangle. However, we do not have enough information to use the Pythagorean theorem. NOT sufficient

(2) BC alone is not enough information since we cannot create an equation for the radius. We are given BC, but we do not know the length of AB. We do not have enough information to write a Pythagorean theorem. NOT sufficient

By taking (1) and (2) together, we can use the Pythagorean theorem for triangle ABD since DA is perpendicular to AB. Let $DC = DA = r$ and let $BD = 18 + r$:

$$(18 + r)^2 = 24^2 + r^2$$
$$324 + 36r + r^2 = 576 + r^2$$
$$36r = 252$$
$$r = 7$$

The answer is C; both statements together are sufficient.

9. **D** Set up a simple proportion to find the number of cookies:

$$\frac{\$}{\text{number}} = \frac{b}{k} = \frac{m}{?} \qquad \text{Cross multiply.}$$
$$? = \frac{mk}{b}$$

The answer is D.

10. **C**

$$\left(\frac{\sqrt{7}}{\sqrt{14}} + \frac{3\sqrt{3}}{\sqrt{6}}\right) \div \frac{\sqrt{16}}{\sqrt{2}}$$

$$= \left(\frac{\sqrt{7}}{\sqrt{2} \cdot \sqrt{7}} + \frac{3\sqrt{3}}{\sqrt{2} \cdot \sqrt{3}}\right) \div \frac{4}{\sqrt{2}} = \left(\frac{1}{\sqrt{2}} + \frac{3}{\sqrt{2}}\right) \cdot \frac{\sqrt{2}}{4}$$

$$= \left(\frac{4}{\sqrt{2}}\right) \cdot \frac{\sqrt{2}}{4} = 1$$

11. **D** (1) The cost of one pair $= \frac{5T}{3}$.

The sales price of one pair $= \frac{6T}{2} = 3T$, so the percent profit:

$$\frac{\text{New} - \text{Original}}{\text{Original}} \cdot 100\% = \frac{3T - \frac{5T}{3}}{\frac{5T}{3}} \cdot 100\%$$

The T's can be factored out in the numerator and cancelled out. So a numerical value can be calculated. SUFFICIENT

(2) Let the cost be x. The sales price becomes $\frac{9x}{5}$. The percent profit is:

$$\frac{\text{New} - \text{Original}}{\text{Original}} \cdot 100\% = \frac{\frac{9x}{5} - x}{x} \cdot 100\%$$

Similar to (1), the x's can be factored out in the numerator and cancelled out. So a numerical value can be calculated. SUFFICIENT

The answer is D; each statement alone is sufficient.

12. **D** Review each answer choice separately:

(A) Since $m < n < 0$, $|m| > |n|$. For example, $-5 < -3 < 0$ and $|-5| > |-3|$. So $m^2 - n^2$ is always positive. TRUE

(B) $\frac{n}{t}$ is negative since $n < 0 < t$ and a negative divided by a positive is always negative. Similarly, $\frac{t}{n}$ is also negative. A negative plus a negative is always negative. TRUE

(C) $\frac{m}{n}$ and $\frac{n}{m}$ are both positive since a negative divided by a negative is always positive. Remember that a positive plus a positive is always positive. TRUE

(D) $mn - mt = m(n - t)$. Since $n < 0 < t$, $n - t$ must be negative. Remember that m is also negative. A negative multiplied by a negative is always positive. FALSE

(E) $nt - nm = n(t - m)$. Since $m < 0 < t$, $t - m$ must be positive. Remember that n is also negative. A positive multiplied by a negative is always negative. TRUE

Once you realize that (D) is the answer, you do not have to review answer choice (E).

13. **B** (1) We can write each variable in terms of one of the others, but that does not give a numerical value. For example, $a = 2b$ and $b = 3c$, which also means $2b = 6c$ and $a = 2b = 6c$.

$\frac{1}{6c} + \frac{1}{3c} + \frac{1}{c}$ cannot be solved for a numerical value. NOT sufficient

(2) $\frac{1}{a} + \frac{1}{b} + \frac{1}{c} = \frac{bc}{abc} + \frac{ac}{abc} + \frac{ab}{abc} = \frac{bc + ac + ab}{abc}$

Since $ab + ac + bc = 2abc$, $\frac{bc + ac + ab}{abc} = \frac{2abc}{abc} = 2$. SUFFICIENT

The answer is B; statement (2) alone is sufficient.

14. **D** If the two given expressions are consecutive even integers, their difference must be 2. However, we do not know which one is the greater one. Try two different scenarios. First assume $t + 8 > 2t - 6$:

$$(t + 8) - (2t - 6) = 2$$
$$t + 8 - 2t + 6 = 2$$
$$t = 12$$

If $t = 12$, the numbers are 18 and 20. The sum must be 38.

Then assume $2t - 6 > t + 8$:

$$(2t - 6) - (t + 8) = 2$$
$$2t - 6 - t - 8 = 2$$
$$t - 14 = 2$$
$$t = 16$$

If $t = 16$, the numbers are 24 and 26. The sum is 50.

The second option gives us a higher sum.

15. **B** Simplify the question first. We are looking for:

$$\frac{a^{-2} - b^{-2}}{a^{-1} - b^{-1}} = \frac{\dfrac{1}{a^2} - \dfrac{1}{b^2}}{\dfrac{1}{a} - \dfrac{1}{b}} = \frac{b^2 - a^2}{a^2 b^2} \cdot \frac{ab}{b - a}$$

$$= \frac{(b - a) \cdot (a + b) \cdot ab}{ab \cdot ab \cdot (b - a)} = \frac{b + a}{ab}$$

(1) This statement provides $\dfrac{a - b}{ab} = -\dfrac{2}{5}$, which does not give us the value of $\dfrac{b + a}{ab}$. NOT sufficient

(2) Since $a + b = \dfrac{3}{5} ab$ is given, divide both sides by ab to get $\dfrac{b + a}{ab} = \dfrac{3}{5}$. SUFFICIENT

The answer is B; statement (2) alone is sufficient.

16. **B** To find the highest possible value, pick the largest $x = 8$ and lowest $y = -3$.

$$3x - 2y = 3(8) - 2(-3) = 30$$

To find the lowest possible value, pick the lowest x. Since $-6 < x$ and x is an integer, $x = -5$. The highest $y = 4$.

$$3x - 2y = 3(-5) - 2(4) = -15 - 8 = -23$$

The maximum minus the minimum is

$$30 - (-23) = 53$$

17. **C** Both roots must satisfy the equation. Plug them in for x and create two equations. Instead of using x_2, plug in x_1 and $3x_1$.

$$x_1^2 - 4x_1 + m - 3 = 0$$

$$(3x_1)^2 - 4 \cdot 3 \cdot x_1 + m - 3 = 0$$
$$9x_1^2 - 12x_1 + m - 3 = 0 \qquad \text{Subtract the first equation}$$
$$\underline{-\ x_1^2 - 4x_1 + m - 3 = 0}$$
$$8x_1^2 - 8x_1 = 0$$
$$8x_1^2(x_1 - 1) = 0$$
$$x_1 = 0 \quad \text{or} \quad x_1 = 1$$

If $x_1 = 1$, then $x_2 = 3$. Plug one of them into the original equation to find m. Note that $x_1 = 0$ does not satisfy the initial conditions:

$$1^2 - 4 \cdot 1 + m - 3 = 0$$
$$m = 6$$

Alternatively, try answer choices until you get two roots where $x_2 = 3x_1$.

18. **D** The question can be translated as follows into $b = x \cdot (a + b)$. So the fraction we are looking for is $\dfrac{b}{a + b}$.

(1) The statement translates into $\dfrac{a + b}{a} = 6$. If you cross multiply, you will get $a + b = 6a$, which means $b = 5a$. So $\dfrac{b}{a + b} = \dfrac{5a}{a + 5a} = \dfrac{5a}{6a} = \dfrac{5}{6}$. SUFFICIENT

(2) a is 20% of b means $a = 0.2b$ or $5a = b$. So, $\dfrac{b}{a + b} = \dfrac{5a}{a + 5a} = \dfrac{5a}{6a} = \dfrac{5}{6}$. SUFFICIENT

The answer is D; each statement alone is sufficient.

19. **B**

$5^{n + 3}$ can be written as $5^n \cdot 5^3 = 5^n \cdot 125$.
$5^{n + 1}$ can be written as $5^n \cdot 5$.

$$125 \cdot 5^n - 5 \cdot 5^n = 3{,}000$$
$$5^n(125 - 5) = 3{,}000$$
$$5^n = \dfrac{3{,}000}{120} = 25 = 5^2$$
$$n = 2$$

20. **D** Translate the first statement:

$\dfrac{1}{3}(a + b) = a - b$ Multiply by 3.

$a + b = 3a - 3b$ Add $3b$ and subtract a from both sides.

$4b = 2a$ Divide both sides by 2.

$2b = a$

Translate the second statement:

$a^2 + b^2 = 245$ Substitute $a = 2b$.

$$(2b)^2 + b^2 = 245$$
$$4b^2 + b^2 = 245$$
$$5b^2 = 245$$
$$b^2 = 49$$

$b = 7$ The question says they are positive integers.
$a = 14$

The difference of squares $= 14^2 - 7^2 = 147$.

21. **B** (1) $a \cdot b < 0$ does not provide enough information. If a is positive, b is negative. If b is negative, a is positive. NOT sufficient

(2) If $a - b > 0$, you could multiply both sides by -1 and flip the inequality sign to get $b - a < 0$. SUFFICIENT

The answer is B; statement (2) alone is sufficient.

22. **D** If the cost of the purse is $120, the retailer must sell it at $144 to make a 20% profit:

$120 \cdot 1.2 = \$144$ or
$120 + 120 \cdot 20\% = 120 + 24 = \$144$

The retailer wants to receive $144 after offering a 20% discount. So the list price less the

20% off the list price must be $144. Let the list price be x:

$$x - 0.2x = 144$$
$$0.8x = 144$$
$$x = \frac{144}{0.8} = \$180$$

23. **D** Let the number of shares of stock X in portfolio A be A. So $A = 2{,}400 - 60d$, where d is the number of days.

Let the number of shares of stock X in portfolio B be B. So $B = 320 + 100d$.

If they become equal in d days, set these two expressions equal to each other:

$$2{,}400 - 60d = 320 + 100d$$
$$2{,}400 = 320 + 160d$$
$$2{,}080 = 160d$$
$$13 = d$$

24. **C** The probability of picking a blue ball in the first pick is

$$\frac{\text{Number of blue}}{\text{Total}} = \frac{k}{k + t}$$

Since there's no replacement, the new total is $k + t - 1$. The probability of picking a red ball in the second pick is

$$\frac{t}{k + t - 1}$$

Picking a blue first and a red second is the product of the two:

$$\frac{k}{k + t} \cdot \frac{t}{k + t - 1}$$

25. **C** (1) We know that $0 < a < 1$. So a is a positive fraction less than 1, but we still do not have any information about b. If $b = 3$ and $a = \frac{1}{2}$, the result would be positive, but $b = \frac{3}{4}$ and $a = \frac{1}{2}$ would result in a negative value. NOT sufficient

(2) $ab > b$ does not provide enough information by itself since we do not know if b is positive or negative. If b is negative, then dividing by b would flip the inequality so a becomes less than 1. If b is positive, the inequality stays the same and a becomes greater than 1. NOT sufficient

By using (1) and (2) together, we can conclude that b is a negative number. When we multiply any positive number by a number between 0 and 1, the number gets smaller.

For example $3 \cdot \frac{1}{2} < 3$ or $\frac{3}{4} \cdot \frac{1}{2} < \frac{3}{4}$. If b is negative, $-3 \cdot \frac{1}{2} > -3$ as in statement (2).

If b is a negative number, $\frac{b - 1}{b - a}$ would always be positive. For example, $\frac{-3 - 1}{-3 - \frac{1}{2}}$.

The answer is C; both statements together are sufficient.

26. **C** (1) $\frac{4}{r} + s = \frac{4}{r} + \frac{rs}{r} = \frac{4 + rs}{r} = 18$. You could also cross multiply. However, $4 + rs = 18r$ does not provide enough information. NOT sufficient

(2) $\frac{4}{s} + r = \frac{4}{s} + \frac{rs}{s} = \frac{4 + rs}{s} = 12$. You could also cross multiply. However, $4 + rs = 12s$ still does not provide enough information. NOT sufficient

By using (1) and (2) together, we can say $4 + rs = 18r$ and $4 + rs = 12s$. Set them equal to each other.

$18r = 12s$ so $\frac{r}{s} = \frac{12}{16}$

The answer is C; both statements together are sufficient.

27. **A** Let the number of boys be x. The number of girls becomes $2x$, and the total players becomes $3x$.

We can calculate the probability of picking two girls as if we are picking them consecutively (obviously without replacement):

$\frac{2x}{3x}$ is picking a girl in the first round.

$\frac{(2x - 1)}{(3x - 1)}$ is picking a girl in the second round.

$\frac{2x}{3x} \cdot \frac{2x - 1}{3x - 1} = \frac{22}{51}$ Simplify the x's on the left.

$\frac{2}{3} \cdot \frac{2x - 1}{3x - 1} = \frac{22}{51}$ Multiply both sides by $\frac{3}{2}$.

$\frac{3}{2} \cdot \frac{2}{3} \cdot \frac{2x - 1}{3x - 1} = \frac{3}{2} \cdot \frac{22}{51} = \frac{11}{17}$

$\frac{2x - 1}{3x - 1} = \frac{11}{17}$ Cross multiply.

$34x - 17 = 33x - 11$

$x = 6$

There are 6 boys and 12 girls.

The probability of picking two boys consecutively is

$\frac{6}{18} \cdot \frac{5}{17} = \frac{5}{51}$

28. **D** (1) Let the number of days Ian worked be x. The number of days he did not work is $(30 - x)$:

$$2,600 = 140x + (30 - x) \cdot (-60)$$
$$2,600 = 140x - 1,800 + 60x$$
$$4,400 = 200x$$
$$22 = x$$

SUFFICIENT

(2) Let the number of days Ian worked be x. The number of days he did not work is $(30 - x)$:

$$x = 6 + 2(30 - x)$$
$$x = 6 + 60 - 2x$$
$$3x = 66$$
$$x = 22$$

SUFFICIENT

The answer is D; each statement alone is sufficient.

29. **B**

Average rate $= \dfrac{\text{Total charge}}{\text{Total hours}}$

Let the time spent by the senior consultant be t hours. The associate spent $2t$ hours, and the associate's rate is $M \cdot \frac{2}{5} = \frac{2m}{5}$.

The total charge is

$$M \cdot t + \frac{2M}{5} \cdot 2t = \frac{5Mt}{5} + \frac{4Mt}{5} = \frac{9Mt}{5}$$

The average rate is $\dfrac{\dfrac{9Mt}{5}}{3t} = \dfrac{9Mt}{5} \cdot \dfrac{1}{3t} = \dfrac{3M}{5}$

30. **E**

$$\text{The total cost} = \text{Cost of first mile} + \text{Cost of the rest}$$

The cost of the first mile is t. For the rest of the trip, use the fee per mile times the miles. The fee per 0.2 miles is $\dfrac{n}{100}$ since n is in cents. The cost per mile is $5 \cdot \dfrac{n}{100} = \dfrac{n}{20}$. Let the total miles we are trying to find be m:

$$P = t + (m - 1) \cdot \frac{n}{20}$$

Notice that we use $(m - 1)$ since the first mile is already paid for by t.

$P = t + (m - 1) \cdot \dfrac{n}{20}$ Subtract t from both sides

$P - t = (m - 1) \cdot \dfrac{n}{20}$ Multiply both sides by $\dfrac{20}{n}$.

$\dfrac{20}{n}(P - t) = m - 1$ Add 1 to both sides.

$\dfrac{20}{n}(P - t) + 1 = m$

31. **D** (1) Let's assume Joe made x on the 11th day. Since on the 12th day the earnings will increase by x, he will make $2x$. Basically, his earnings double every day.

11th day: x
12th day: $2x$
13th day: $4x$
14th day: $8x$

Since $8x = \$573.44$, x can be easily found. SUFFICIENT

(2) Let's assume Joe made y on the first day:

1st day: y
2nd day: $2y$
3rd day: $4y$
4th day: $8y$
5th day: $16y$
6th day: $32y$

The sum for the 6 days is $y + 2y + 4y + 8y + 16y + 32y = \$4.41$. This can be easily solved for y. SUFFICIENT

The answer is D; each statement alone is sufficient.

32. **A** If Ronald makes $1,980 in 3 years at simple interest, he makes $\frac{\$1,980}{3} = \$660$ per year.

Principal × Interest rate = Interest ($)

	$P	0.06	0.06P
Inv. 1	$P	0.06	0.06P
Inv. 2	$10,000 − P	0.075	(10,000 − P)0.075
Total			660

$\text{Interest}_1 + \text{Interest}_2 = \text{Total interest}$
$0.06P + 0.075(10,000 − P) = 660$
$0.06P + 750 − 0.075P = 660$

$90 = 0.015P$
$P = \$6,000$ at 6%
So he earned $4,000 at 7.5%.
The ratio is $\frac{6,000}{4,000} = \frac{3}{2}$.

33. **C** Let the number of e-mails to be sent be n. The number of e-mails that reach inboxes is $60\% \cdot n = 0.6 \cdot n$. Only 20% of the people click the links. So $0.2 \cdot 0.6 \cdot n$ people go to the website. Finally, $\frac{1}{3}(0.2 \cdot 0.6 \cdot n)$ surveys get completed.

$\frac{1}{3}(0.2 \cdot 0.6 \cdot n) = 360$

$n = 9,000$ e-mails

34. **C** The x-coordinate of the center must be n since the x-coordinate of point B is n. Furthermore, since the circle is tangent to the y-axis at A, its radius is n. So $OA = OB = OD = n$.

If the y-coordinate of point B is m, then m must be a negative number. The distance from the center to the x-axis is $n + m$. Finally, since $OD = n$, the y-coordinate of D must be $n + n + m = 2n + m$.

So the coordinates of point D are $(n, 2n + m)$.

35. **D** Cross multiply the question. We are looking for $kn = mt$.

(1) Cross multiply to get $k(n − m) = m(t − k)$.
$$kn − km = mt − mk$$
$$kn = mt \qquad \text{SUFFICIENT}$$

(2) Cross multiply to get $k(n + m) = m(t + k)$.
$$kn + km = mt + mk$$
$$kn = mt \qquad \text{SUFFICIENT}$$

The answer is D; each statement alone is sufficient.

36. **C** The length of the diagonal of $ABCD$ is $\sqrt{x^2 + y^2} = d_1$.

The length of the diagonal of $EFGH$ is $\sqrt{(x+y)^2 + (x-y)^2} = d_2$.

$$d_2 = \sqrt{x^2 + 2xy + y^2 + x^2 - 2xy + y^2}$$
$$= \sqrt{2x^2 + 2y^2} = \sqrt{2} \cdot \sqrt{x^2 + y^2}$$
$$\frac{d_2}{d_1} = \frac{\sqrt{2} \cdot \sqrt{x^2 + y^2}}{\sqrt{x^2 + y^2}} = \sqrt{2}$$

37. **B** Simplify the question first:

$$\frac{mn}{m.n} = 10 \quad \text{and} \quad \frac{mn}{0.mn} = 100$$

$\frac{mm}{10}$ cannot be simplified without knowing m.

(1) $n = 2$ is not enough since all we need is m. NOT sufficient

(2) $m = 3$ is enough information since $\frac{mm}{10} = \frac{33}{10} = 3.3$. SUFFICIENT

The answer is B; statement (2) alone is sufficient.

INDEX

Integers, 19–23, 139
Intercepts, 224
Interior angle, 183
Inverse proportions, 51–52
Investment/interest word
 problems, 168–172
Irrational numbers, 19
Isosceles triangle, 186

Least common multiple, 26
Legs, of triangle, 184
Linear equations, 99–102,
 222–223
Linear inequalities, 103–104
Lines
 equation of a, 225
 in graphs, 222–223
 types of, 181–182

Median, 67
Midpoints, 220
Mixed numbers, 30
Mixture problems, 161–168
Mode, 67
Monomial, 81
Multiples, 24
Multiplication
 of algebraic expressions, 92
 of decimals, 36
 of exponents, 110–111
 of fractions, 31
 of radicals, 112–113
 of signed numbers, 21
Mutually exclusive events, 78

Negative exponents, 111
nth term, 124
Numbers
 irrational, 19
 mixed, 30
 prime, 21
 rational, 19
 real, 19–20
 signed, 21–22

Odd integers, 20–21
Ordered pair, 219, 221
Ordering, 75
Order of operations, 22
Origin, 219

Parabolas, 225–226
Parallel lines, 182, 225
Parallelograms, 200–201
Partial investments, 169
Percentages, 58–59
Perfect squares, 12

Perimeter, 183
Permutation, 75
Perpendicular lines, 182, 225
Pi, 202
Place value, 27
Polygon, 182–183, 204
Polynomial, 82
Preparation strategies, 9–12
Prime factorization, 25–26
Prime numbers, 21
Probability
 discrete, 76–77
 of two or more events, 77–78
Problem-solving questions, 5
Product of 0, 121
Proportions, 50–52
Pythagorean theorem, 184,
 201, 220

Quadratic equations, 120–122
Quadratic formula, 121
Quadratic functions, 225–226
Quadrilaterals, 200–204
Quantitative section, 15–18
Questions
 data sufficiency. *See* Data
  sufficiency questions
 experimental, 6
 problem-solving, 5

Radicals, 111–113
Radius, 202
Range, 67, 123
Rate problems, 148–155
Ratio
 common, 124
 description of, 50–51
Rational numbers, 19
Ratio of the areas, 203
Real numbers, 19–20
Reciprocal, 22
Rectangles, 201
Rectangular coordinate
 system, 219
Rectangular prisms, 204–205
Reducing fractions, 30
Registering, 18
Remainders, 26
Repeating sequences, 125
Right angle, 182
Right triangle, 184–186
Roots of quadratic
 equations, 122
Rounding, 27

Scores, 1, 3, 6–7, 11–12

Sequences, 124–125
Set problems, 173–179
Side, 182
Signed numbers, 21–22
Similar triangle, 187
Simple interest, 168
Simplifying radicals, 112
Slope, 223–224
Solids, 204–205
Solution of equation, 99
Square, 201
Square root, 111, 113
Standard deviation, 67–68
Study plan, 9–10
Substitution, 94, 101
Subtraction
 of algebraic expressions, 92
 of decimals, 36
 of fractions, 31
 of radicals, 112
 of signed numbers, 22
Sum of all interior angles, 183, 187
Sum of n terms, 124
Symbolism, 123

Table, 234–236
Tangent, 202
Term, 81, 124
Test-day strategies, 12–15
30-60-90 triangle, 185
3-4-5 triangle, 184
Trapezoids, 201–202
Triangles, 184–188

Value questions, 243–244
Variables, 81
Venn diagrams, 173–179
Vertex, 182, 204
Volume, 205

Word problems, 135–180
 age, 138
 exchange, 138
 investment/interest, 168–172
 mixture, 161–168
 rate, 148–155
 set, 173–179
 solving, 138–139
 translation to equations,
  135–148
 work, 156–161
Work problems, 156–161

x-intercept, 224

Yes/no questions, 245–246
y-intercept, 224